THE UNITED STATES AIR FORCE IN BRITAIN

DARREN WILLMIN

FONTHILL

Fonthill Media Language Policy

Fonthill Media publishes in the international English language market. One language edition is published worldwide. As there are minor differences in spelling and presentation, especially with regard to American English and British English, a policy is necessary to define which form of English to use. The Fonthill Policy is to use the form of English native to the author. Darren Willmin was born and educated in England; therefore British English has been adopted in this publication.

Fonthill Media Limited
Fonthill Media LLC
www.fonthillmedia.com
office@fonthillmedia.com

First published in the United Kingdom and the United States of America 2019

British Library Cataloguing in Publication Data:
A catalogue record for this book is available from the British Library

ISBN 978-1-78155-699-3

Typeset in 10pt on 13pt Sabon
Printed and bound in England

Acknowledgements

48th Fighter Wing: TSgt (Technical Sergeant) E. Burks (Public Affairs); Capt. J. Ford (493rd Pilot); Capt. A. Eshkenazi (493rd Pilot); Capt. M. 'Switcha' Rioux (493rd Pilot); Capt. B. Davis (493rd Pilot); Col. Evan L. Pettus (Commander of 48th Fighter Wing); and all other personnel who were involved in the book.

100th ARW: SMSgt (Senior master sergeant) B. Boisvert (Public Affairs); S/Sgt K. Thornton (Public Affairs); SrA (Senior Airman) J. Rho (Public Affairs); Capt. J. Fuller (100 ARW Pilot); Capt. R. Barnes (100 ARW Pilot); SrA (Senior Airman) K. Bryant (100 ARW Boom Operator); Col. Christopher Amrhein (100 ARW Base Commander); and all other personnel who were involved in the book.

501st CSW: Capt. A. McGee (Public Relations); 1Lt K. Jimenez (Public Affairs).

352nd Public Relations; 56RQS Public Relations; 52nd Fighter Wing Spangdahlem AFB (Public Relations); 388th Fighter Wing, Hill AFB (Public Relations); 325th Fighter Wing, Tyndall AFB (Public Relations); SrA S Fales 5th Expeditionary Operations Group, Minot AFB (Public Affairs); Col. K. Denehan (Group Commander); Maj. F. Devita (Support Squadron Commander); Lt-Col. M. Maginness (23rd Expeditionary Bomb Squadron Commander); Chief Turner (5th Maintenance Superintendent); 1Lt M. Roux (USAFR Deputy Chief, 307th Bomb Wing) Barksdale AFB (Public Affairs & Protocol); TSgt N. Filip (Public Affairs and Protocol); Col. Denis Heinz 489th Bomb Group, Dyess AFB, Texas (Commander); Capt. J. Bresnahan 11th Bomb Squadron, Barksdale AFB (Pilot); Capt. B. Kerns (Instructor Electronic Warfare Officer); Capt. D. Skelonc (Instructor Weapons System Operator); Maj. C. Platner 93rd Bomb Squadron, Barksdale AFB (Instructor Pilot); Lt-Col. B. McClanahan 345th Bomb Squadron, Dyess AFB (Commander); 9th Reconnaissance Wing Public Affairs; Seymour Johnson Public Affairs; and 509th Bomb Wing, Whitman AFB, Public Affairs.

Contents

Glossary and Abbreviations

1LT:	First Lieutenant
AB:	Air Base
ABG:	Air Base Group
ABS:	Air Base Squadron
ABW:	Air Base Wing
ACC:	Air Combat Command
AD:	Air Division
AFB:	Air Force Base
AMC:	Air Mobility Command
ARS:	Air Refuelling Squadron
ARW:	Air Refuelling Wing
ATC:	Air Traffic Control
BG:	Bombardment Group
BS:	Bomb Squadron
BW:	Bomb Wing
Capt.:	Captain
CAS:	Close Air Support
Col.:	Colonel
CSW:	Combat Support Wing
EATS:	European Air Transport Service
ECS:	Electronic Combat Squadron
EOR:	End of Runway
FEW:	Fighter-Escort Wing
FIW:	Fighter Interceptor Wing
FW:	Fighter Wing
HQ:	Headquarters
JTAC:	Joint Terminal Attack Controllers
LFA:	Low Flying Area
Maj.:	Major
MAJCOM:	Major Command
MDP:	Ministry of Defence Police

MOD:	Ministry of Defence
NATO:	North Atlantic Treaty Organization
PAS:	Permanent Aircraft Shelter
PR:	Public Relations
RAF:	Royal Air Force
RAPCON:	Radar Approach Control
RIAT:	Royal International Air Tattoo
RQS:	Rescue Squadron
RW:	Reconnaissance Wing
SAC:	Strategic Air Command
SOW:	Special Operations Wing
SrA:	Senior Airman
SRW:	Strategic Reconnaissance Wing
SSgt:	Staff Sergeant
TCW:	Tactical Combat Wing
TDA:	Temporary Duty Assignment
TDY:	Temporary Duty
TFW:	Tactical Fighter Wing
TRS:	Tactical Reconnaissance Squadrons
TRW:	Tactical Reconnaissance Wing
TSgt:	Technical Sergeant
USAAF:	United States Army Air Forces
USAF:	United States Air Force
USAFE:	United States Air Forces in Europe
WSO:	Weapon Systems Officer

Introduction: The United States Air Force History

USA and UK Flag and USAFE GB

The United Kingdom and the United States governments signed a series of agreements in the early part of 1941, which allowed the American Naval, Army, and Air Forces to design and plan campaigns against Nazi Germany.

Since 1942, the United States have maintained air bases and major command centres of the United States Air Force in the United Kingdom; these are known as the United States Air Forces in Europe (USAFE), Strategic Air Command (SAC), and Air Mobility Command (AMC).

With the end of the Second World War, the US began to demobilise a majority of the Air Force that it had created in the United Kingdom.

In Europe, the aim was to keep a small United States Army Air Force, predominantly for communication and transport purposes, and on 7 August 1945, the word 'Strategic' was removed from USSAFE and the United States Air Forces in Europe (USAFE) was established. By the end of 1946, USAFE had only about 75,000 personnel and fewer than 2,000 aircraft.

Today in the UK, the USAF has three operational flying airbases, one being RAF Mildenhall, located near the town of Mildenhall in Suffolk. The host unit at Mildenhall is the 100th Air Refuelling Wing (100 ARW) operating the Boeing KC-135 Stratotanker, which deploys aircraft for and manages the European Tanker Task Force. It also houses the 352 SOW, which is the air force component of the Special Operations Command in Europe, who operates the C130J series and the CV-22 Osprey—a sub-unified command of the US European Command. The 95th Reconnaissance Squadron, which conducts RC-135 Rivet joint flight operations in Europe and Mediterranean theatres, uses RAF Mildenhall as a secondary base.

RAF Lakenheath, located near the town of Lakenheath in Suffolk, is one of the other airbases and hosts the 48th Fighter Wing (48 FW), also known as the Liberty Wing, assigned to United States Air Forces in Europe–Air Forces Africa. The 48th Fighter Wing aircraft is the McDonnell Douglas (now Boeing) F-15 Eagle, and here it houses three squadrons plus the 56th Rescue Squadron (56 RQS) as part of the wing. The 56 RQS are a combat-ready search and rescue squadron of HH-60G Pave Hawk helicopters capable of executing all-weather search and rescue missions in support of USAFE.

RAF Fairford in Gloucestershire, England, is the third location and is currently a standby airbase and therefore not used every day. In recent years, and even to this day, it has been used as an airfield for the USAF and was used by B-52s during the First Gulf War in 1991, Operation Allied Force in 1999, and the 2003 Iraq War, and it remains the USAF's only European airfield for heavy bombers.

USAF History

USAFE–AFAFRICA is one of the major commands, and a component of the United States European and the United States Africa Commands.

As part of USAFE, it controls the US forces' units pledged to NATO, in which it maintains combat-ready wings based in the UK across to Turkey.

USAFE–AFAFRICA coordinates, controls, and conducts air support and operations in Europe and Africa, except for Egypt. The headquarters for USAFE–AFAFRICA is at Ramstein Air Base in Germany, which is the oldest active Major Command.

The command was activated in 1942 at Langley Field, Virginia, as the Eighth Air Force by the USAAF. In 2012, USAFE became USAFE–AFAFRICA. The command has more than 35,000 active personnel, consisting of air reserve and civilian personnel.

During the mid- to late 1940s, USAFE was occupied with supporting the movement of men and aircraft of the Strategic Air Command (SAC) to bases in England.

USAFE originated as the USAAF Eighth Air Force back in 1942. The Eighth Air Force commanded 8th Bomber, 8th Fighter, and 8th Air Support Command. In 1944, the Eighth Air Force was redesignated as the United States Strategic Air Force in Europe (USSTAF) and the 8th Bomber Command, which was redesignated as the Eighth Air Force, was assigned to the control of USSTAF. In 1945, the USSTAF was redesignated as USAFE, with its headquarters relocated from Saint-Germain-en-Laye in France to Lindsey Air Station, Wiesbaden, in Germany.

At the end of the Second World War, virtually all the US personnel had left Europe, apart from the occupation forces in Germany and Austria. Additionally, USAFE personnel had been reduced from over 500,000 to 77,000. In 1945, USAFE was given the status of a major command (MAJCOM).

The European Air Transport Service (EATS) provided passenger and cargo transport within Western Europe using the C-146, C-47, and C-54 aircraft, with their headquarters based in Wiesbaden, Germany. EATS had bases in many European countries, including Germany, Italy, and Austria.

During 1946, tensions began to rise with the Soviet Union, and President Truman decided to realign USAFE to become combat ready. In November 1946, six B-17 Superfortress bombers from the 43rd Bombardment Group arrived at RAF Burtonwood from West Germany on a training deployment.

With activity from the Soviet Union growing, it started to unsettle the western allies, so President Harry S. Truman decided to take a hard approach with Russia, which evolved into the Cold War. In Germany, three air bases were rebuilt to accommodate the Boeing B-29 Superfortress bombers. The Strategic Air Command (SAC) wanted the B-29s as close as possible to the Soviet Union's border because of their limited range on fuel. Many B-29s were flown to bases in France, Turkey, and Greece, and they often flew along the borders of Bulgaria and Russia. More B-29s were sent to the UK in 1947 to keep up the presence, and

A formation of Second World War aircraft grace the skies over Duxford, including the B-17 Flying Fortress and P-51 Mustang.

these deployments were only a cover up of the true aim, which was to have these B-29s as a strategic air force permanently stationed in Europe.

All B-29 operations in the UK were placed under the command of the newly formed 3rd Air Division based at RAF Bomber Command at RAF Marham; additionally, the RAF had several airfields at their disposal, which included RAF Scampton, RAF Waddington, and RAF Lakenheath. One of the first groups to arrive in the UK for a ninety-day tour was the 28th Bombardment Group from Rapid City Air Force Base, South Dakota, and they arrived at RAF Scampton during 1948.

The aircraft from the 28th Bombardment Group had a special designation on both sides of the tail fin; the scheme was a big black letter 'R', which denoted the groups, within a circle, which identified the parent unit: the 15th Air Force. Many units later carried their parent numbered air force unit badge on the base of the tail fin.

The 2nd Air Force carried a black square on its B-29s, and the Eighth Air Force used a triangle. It is believed that when a group or wing transferred from one numbered air force to another, their identification letter remained the same; however, the numbered air force geometric identification symbol changed.

The B-29s often carried colourful personal marking and squadron colours and some B-29s retained the Korean War flat black under surfaces, while others carried red tail sections and outer wing panels known as arctic markings, which were used to make the aircraft more visible in the event of a forced-landing.

During 1949, the B-50 Superfortress from the 43rd Bombardment Group deployed to the UK and carried the style of markings used by the B-29s. In January 1951, the first B-36D was deployed during an exercise (Operation UK) to the UK from the 7th Bombardment Group, who were based at RAF Lakenheath. During 1951 and 1959, B-36s made regular visits to the UK, and the 42nd Bombardment Wing visited RAF Upper Heyford and RAF Burtonwood. In 1954, the entire wing deployed to RAF Heyford.

The Berlin Airlift was one of the events that marked the beginning of the Cold War; between 1947 and 1991, the Cold War was a state of political and military tension between the Western Bloc (The US, its NATO allies, and others) and the powers in the Eastern Bloc (the

Soviet Union and the Warsaw Pact). The airlift was a massive humanitarian effort and was an early triumph for Allied air forces, and it symbolised Western commitment to rebuilding democracy in Europe after the Second World War.

The Cold War split the temporary wartime alliance against Nazi Germany, leaving the USSR and the US as two superpowers with profound economic and political differences; the disagreement was about how the European map should look and how the borders would be drawn after the Second World War.

The Berlin Blockade was a major crisis of the Cold War, and during the multinational occupation of Germany, the Soviet Union blocked the Western Allies' access to railway, roads, and rivers in sectors of western-controlled Berlin. The Western Allies organised the Berlin Airlift to carry supplies to the people of the West Berlin.

Priority was given to supplying Berlin by air, as the Soviet blockade had insignificant effect on the three Berlin air corridors. The Soviet Union did not initially interfere with the cargo aircraft flying the Berlin Airlift as they were convinced that supplying the population of 2 million by air was an impossible task. The command consisted of 485 aircraft, with the 60th and 61st Troop Carrier Groups at Rhein-Main and Wiesbaden Air Bases near Frankfurt both flying C-47s. The only other flying unit was the 86th Fighter Group at Neubiberg Air Base near Munich, with P-47s.

Aircrews from the US Air Force, the RAF, the Royal Canadian Air Force (RCAF), the Royal Australian Air Force (RAAF), the Royal New Zealand Air Force (RNZAF), and the South African Air Force flew over 200,000 flights in just one year, providing the Berlin people with up to 9,000 tons of supplies each day. It was clear to the Soviets after a few months that the Americans were succeeding in supplying the Western areas of Berlin with the minimal supplies needed. Mock attacks from Soviet fighters were begun to try and scare American pilots. Many hundreds of pilots and many thousands of military and German civilians involved in the airlift kept the people of Berlin supplied.

A rare formation of three Douglas C-47s flypast during Duxford Air Show.

In 1961, Berlin became USAFE's first real test when the Soviet Union and East Germany signed a peace treaty that, in effect, gave the German Democratic Republic control of the Russian zone of Berlin and ended the joint occupation of the city. This was a violation of the Potsdam Agreement of 1945. The Western Allies objected to this proposed peace treaty, while the Soviet-backed East German government began restricting the West's aerial access to Berlin and preventing the entry of East Germans into the city. By the middle of August, over 5,000 refugees were trying to leave Berlin, and on the night of 12 August, the Soviet-backed East German government began to erect the Berlin Wall. By the end of 1961, Berlin had become a divided city, while the response from the US government was to increase its tactical airpower in Europe.

The USAF responded by sending reinforcements to Europe, which led to the largest deployment of overseas aircraft movement since the Second World War. The first phase began with Operation Tack Hammer. During this, Tactical Air Command launched eight F-100D squadrons from its Composite Air Strike Force to strengthen USAFE presence with 144 fighters. All Tack Hammer fighters moved across the Atlantic Ocean with aerial refuelling *en route*. The Air National Guard was tasked to supply tactical fighter wings and one tactical reconnaissance wing to expand USAFE.

The second phase began with the movement of eleven Air National Guard squadrons in late 1961. Operation Stair Step was the codename for the rapid aerial movement of the fighters to Europe. Aircraft supplied by ANG wings were the F-84Es, RF-84Fs, F-86Hs, and F-104As, but the F-84E and F-86Fs were considered a very old aircraft even though they were only seven to nine years old.

The primary combat mission was air superiority and offensive tactical air support to defend West Germany if a war developed over access to Berlin. By 1962, the Berlin Crisis was subsiding, and plans were being made for departure of the ANG wings from Europe. Units were to return to the States, with all personnel and aircraft by September 1962 for initial release from active duty.

The Berlin Wall was built, with a barbed-wire fence and minefields stretching along the entire north and south length of the divided Germany. The wall effectively isolated East Germany for the next two decades, but the American, British, and French Zones remained in Berlin and access to the city was not challenged again. The USAFE forces had served their purpose: their rapid deployment to France had unequivocally demonstrated the United States' determination to defend Berlin.

In 1966, French President Charles de Gaulle announced that France would withdraw from NATO's integrated military structure and gave the NATO forces one year to leave France. During 1967, all USAF offices and facilities in France were closed and personnel and equipment moved to other NATO countries. The last active USAFE force was the 1630th Air Base Squadron at Orly Airport and the Paris Administration Office. A C-47 variant was the last USAF aircraft to depart France on 31 May 1967. With the French departure, a major reorganisation of USAFE was needed. The 49th Tactical Fighter Wing's (TFW) three squadrons at Spangdahlem Air Base and the 417th TFS of the 50th TFW at Hahn Air Base were recalled to the US. Although the squadrons were relocated to the US, they were still part of USAFE's permanent force. According to the Department of Defense, these squadrons were 'dual-based'—they could return to their European bases at any given moment without lengthy preparations being necessary.

North American F-86A Sabre FU-178/8178 beautifully showing the top side of the aircraft.

Above: F-4F Phantom II - 38+28 Squadron *Jagdgeschwader* 71 from the German Air Force departs RAF Waddington, 2012.

Below: Lockheed F-104G Starfighter FX-61 Belgian Air Force based at Kleine Brogel Air Base.

During 1967, the 49th TFW's three squadrons flew back to the US where it was stationed at Holloman Air Force Base, New Mexico. In 1968, the four squadrons switched over to McDonnell Douglas F-4 Phantom fighter-bombers and then undertook intensive preparations for their new role within USAFE. Air Force units from the US mainland would be deployed to Europe for Allied Forces Central Europe exercises. Most of the heavy equipment, such as armoured vehicles, artillery, etc., was shipped by sea to exercise that transportation component. Troops were flown via military and contract transport aircraft.

It took until the mid-1970s before USAFE had fully realigned its forces in Europe. Zweibrücken AB in West Germany and RAF Upper Heyford in England came under USAFE control within the next several years. Older reconnaissance and fighter aircraft were redeployed from France to Southeast Asia to supplement the US Pacific Air Forces engaged in the Vietnam War.

Spain joined NATO in 1982, but before they did so, the USAF had for many years used Spanish air bases. Initially used by the Strategic Air Command, these were Morón Air Base at Morón, near Sevilla in southern Spain, and Torrejón Air Base at Torrejón, near Madrid. The Spanish air bases were important for reinforcing USAFE via the southern Atlantic route. Aircraft that flew to Europe via Lajes Field in the Azores always made a refuelling stop at Morón, and later at Torrejón as well. These bases also have American facilities for carrying out aircraft maintenance and repairs. Yet it was primarily the pleasant weather that drew USAFE to Spain for weapons training.

When SAC's 65th Air Division was transferred to USAFE, the footprint of USAFE's activities in Spain increased with two interceptor squadrons (equipped with F-102 Delta Daggers) being formed: the 431st Fighter-Interceptor Squadron (431 FIS) being stationed at Zaragoza Air Base and the 497th FIS at Torrejon AB. As compensation for the permanent use of these Spanish bases, the CASA aircraft factory at Morón AB was brought in to maintain the F-102A air defence fighters that the USAFE had stationed in Spain, Germany, and the Netherlands.

In the mid-1960s, American–Libyan relations worsened and a growing number of USAFE fighter-bomber squadrons in England and Germany went to Zaragoza and gunnery ranges in Spain for weapons training. Zaragoza later became an important weapons training site for the USAFE and was also visited by F-15 Eagle squadrons for 'Dissimilar Air Combat Training'. During these air combat training exercises, the F-15s often practiced against Spanish Air Force Dassault Mirage F-1 fighters.

In April 1966, the 16th Air Force was transferred from SAC to the USAFE, with USAFE taking control of the Spanish air bases at Zaragoza and Morón. Under USAFE, the Spanish bases became host to a growing number of deployments. Morón received regular visits from Lockheed F-104C Starfighters of the 479th TFW from George AFB, California. During the Cuban Missile Crisis, a squadron of F-104Cs was stationed at Morón, and during the mid-1960s, the 16th Air Force also gradually took over responsibility for all USAFE operations around the Mediterranean.

The US Logistics Group was the primary USAF agency in Turkey. TUSLOG not only commanded various USAFE units, but also supported all other US military organisations and government agencies in Turkey and was headquartered in the Turkish capital of Ankara. The 39th Air Base Wing at Incirlik Air Base supported training deployments and regional exercises, plus communications for National Command Authority undertakings. It also provided support for various units and an Air Mobility Command tenant unit provided air transport of passengers and cargo. From the 1950s to the 1970s, the 39th supported various SAC activities in Turkey, which used Incirlik intensively as a base for U-2 reconnaissance flights along the Soviet border and in the Middle East.

In Ankara, the 7217th Air Base Group managed the logistical support for more than forty units and agencies, as well as the needs of the American Embassy and US Defense *Attaché* Office. From Izmir Air Station, the 7266th Air Base Group supported the two NATO headquarters: Allied Land Forces South-Eastern Europe and the Sixth Allied Tactical Air Force. The 7241st Air Base Group was the only US military unit in Turkey not located at a single site but was scattered about İzmir in various locations. Cutbacks to the forces in Turkey naturally had a major effect on TUSLOG. The headquarters in Ankara shrank to a fraction of its former size, and in 1970, it was deactivated as the 7217th Air Division and the next day re-established as Detachment 1 of Headquarters, Sixteen Air Force.

In the early 1970s, the headquarters of USAFE transferred from Lindsey Air Station, Germany, to Ramstein Air Base, and NATO's Allied Air Forces Central Europe was established at Ramstein Air Base in 1974. The USAFE commander-in-chief then took command of Allied Air Forces Central Europe, in addition to commanding USAF units in Europe.

In 1976, the new McDonnell Douglas F-15A Eagle air superiority fighter was introduced into USAFE service. The Soviet Union's new MiG and Sukhoi fighters made the US Department of Defense anxious. The MiG-25 'Foxbat' made them pull out all the stops to get the F-15A into USAFE. The F-15A was deployed to Germany in 1977 with the 36th TFW at Bitburg Air Base West Germany. The 32nd TFS at Soesterberg AB Netherlands was also upgraded to the McDonnell-Douglas F-15A Eagle as part of Project Ready Eagle, and by 1986, all USAFE F-4 wings were replaced by F-15 and F-16 fighters. However, after flying the F15A and F-15B for just eighteen months, the USAFE exchanged these models for the newer F-15C and F-15D Eagles.

With the Iraq crisis in the 1990s, Operations Desert Shield and Desert Storm 1991, more than 180 aircraft and 5,400 personnel assigned to USAFE units deployed to the Persian Gulf area. The command's air support was lethal, with the USAFE accounting for only 20 per cent of the air-to-air assets in Desert Storm, claiming half of the air-to-air kills and more than 85,000 tons of munitions, including more than 35,000 bombs and 7,800 missiles in the theatre. These were used in countless strikes, interdictions, and close air support missions.

After Desert Storm ended, Kurdish rebels and Iraqi forces continued fighting in northern Iraq. The Kurds began a mass exodus towards Turkey and later Iran. A multi-national effort, including US forces, was slowly established to save lives during Operation Provide Comfort. The operation immediately began air dropping food and supplies to the refugees. More than 2,400 USAFE people were deployed, along with thirty-six fighter aircraft, to provide protection for the transports.

The original mandate from the Turkish government allowed the operation to continue for six months. Turkey subsequently approved two six-month extensions, but indicated that it would not become a permanent mission. For the first year of the mission, northern Iraq was quiet, with no combat between coalition aircraft and Iraqi forces.

The US and coalition aircraft over northern Iraq came under almost daily fire from Iraqi surface-to-air missile sites and anti-aircraft guns. These aircraft responded by bombing Iraqi air-defence sites, which fired on them, utilising laser-guided bombs as well as AGM-88 HARM

F-15C Strike Eagle 84-0044 from 493rd Fighter Squadron from RAF Lakenheath taxies to parking at Kleine Brogel Spotters day.

Two F-15Cs Strike Eagles from the 493rd Fighter Squadron break for landing at RAF Lakenheath.

missiles and AGM-130 long range air-to-surface missiles. Coalition aircraft flew patrols on an average of eighteen days per month and were usually fired upon. The most common threat was from anti-aircraft guns. Despite Saddam Hussein offering a reward for downing a coalition aircraft, no warplanes were ever shot down.

The conflict over northern Iraq continued up until the invasion of Iraq in 2003, although the number of incidents declined dramatically after 1999. The final combat air patrol occurred on March 2003 from Incirlik Air Base. Six weeks later, the operation concluded with an official stand down on 1 May 2003. A total of 36,000 sorties were flown during Operation Northern Watch, and 40,000 personnel had been deployed at some point during the operation.

USAFE also provided air protection over the skies of Bosnia-Herzegovina in Operation Deny Flight. Along with allies from NATO countries, US aircrews bombed targets in Bosnia-Herzegovina during Operation Deliberate Force, which paved the way for the 1995 Dayton Peace Agreement. USAFE then helped deploy the Implementation Force and its equipment to Bosnia for Operation Joint Endeavour and sustained them by airlift.

USAFE forces again mobilised in March 1999, when NATO intervened in Kosovo to halt a Yugoslav counter-insurgency targeting the Kosovo Liberation Army. USAFE forces provided air-support for Albanian fighters on the ground. Albanian refugees appeared after the beginning of hostilities. Efforts to find a diplomatic solution collapsed, resulting in Operation Allied Force—the NATO-led air war over Kosovo. The seventy-eight-day operation ended on 20 June, culminating in the withdrawal of Serb forces from Kosovo and the eventual return of refugees. USAFE's Third Air Force led Joint Task Force Shining Hope, established to assist the hundreds of thousands of refugees who left Kosovo because of war. USAFE continues to contribute to NATO-led forces promoting peace and stability in Kosovo.

The War in Afghanistan is the most recent conflict where the USAFE has provided support: an air bridge from Europe to Asia that delivered 3,300 tons of humanitarian daily rations to northern Afghanistan and established a medical evacuation network that moved nearly 4,000 patients. USAFE deployed fighter aircraft, KC-135 Stratotankers, and nearly 2,400 people in Operation Iraqi Freedom. It opened an important airfield in northern Iraq and provided critical *en-route* support to deploying forces, not to mention vital logistical and medical support to forward-deployed forces.

Today, USAFE airmen are engaged in a wide range of active US military efforts in Europe and Africa, including realistic US and NATO exercises and operations in Afghanistan, Iraq, and Syria. In April 2012, US Air Forces Africa AFAFRICA was merged with USAFE to become United States Air Forces in Europe–Air Forces Africa (USAFE-AFAFRICA). The merger was a result of Seventeenth Air Force at Ramstein Air Base deactivating in April 2012 as part of air force cost saving efforts. In 2015, the Department of Defense announced the closure of RAF Mildenhall and the absorption of 700 personnel into Ramstein and Spangdahlem air bases as part of a program to save $500 million a year across Europe. The changes are a result of the US Department of Defense's European Infrastructure Consolidation (EIC), a two-year-long study to downsize the number of sites and facilities being used by Unites States forces in Europe. After the recent presidential elections, when Donald Trump was elected, the military budget is being re-evaluated.

1
Past Operational and Non-Flying Bases

MDPGA Wethersfield is a Ministry of Defence facility in Essex, England, located north of the village of Wethersfield, north-west of Braintree. Once a military air station, the existing accommodation areas have been converted to become the headquarters and training centre of the Ministry of Defence Police.

During the Second World War, RAF Wethersfield was used by the RAF and the USAAF. RAF Wethersfield was closed in 1946 but was reopened in 1951 due to the Cold War. The USAF used the site until 1970 as a fighter base, but it was held as a reserve airfield until 1993 when it finally fell under the control of the Ministry of Defence Police.

RAF Wethersfield was assigned to the USAAF in 1942, and for security reasons, it was known as USAAF-170 rather than the normal location name during the war. The 416th Bombardment Group arrived at RAF Wethersfield in early 1944 with four squadrons all flying the Douglas A-20 'Havoc' aircraft. The group returned to the US and was deactivated at 1945.

The British government provided RAF Wethersfield to the USAF because of the Cold War threat from the Soviet Union and due to the US's commitment to NATO. In 1952, the 20th Fighter-Bomber Wing (FBW) was transferred from Langley AFB in Virginia and took up occupancy at RAF Wethersfield. The 20th FBW consisted of three operational squadrons flying the F-84G 'Thunderjets'. The F-84Gs were specially equipped to carry small nuclear bombs and could deliver these weapons on Soviet forces if an invasion of West Germany was to take place. Due to restricted space at RAF Wethersfield, it was decided that the 79th Squadron was to be stationed at RAF Bentwaters, then at RAF Woodbridge, from 1954. In 1955, the 20th FBW started receiving the F-84F 'Thunderstreak' in addition to its F-84Ds and F-84Gs. The F-84Gs were withdrawn and transferred to Allied nations in Europe and the Middle East.

In 1957, the F-84F was withdrawn and the conversion to the North American F-100D and F-100F 'Super Sabres' began. The F-100 remained the primary aircraft at RAF Wethersfield until 1970.

As part of the air force worldwide reorganisation, the 20th FBW began realigning its units in 1957. As part of another organisation change, the 20th dropped the 'Fighter-Bomber' designation in 1958 and became the 20th Tactical Fighter Wing. The three flying units also changed designation, becoming tactical fighter squadrons. The squadrons were distributed on a monthly rotational basis to RAF Alconbury, RAF Woodbridge, and Nouasseur Air Base in Morocco, due to runway repairs at Wethersfield during the summer of 1958.

Belgium Air Force Republic F-84F Thunderstreak FU-66 housed at Kleine Brogel Air Base.

The 20th began monthly rotations of its fighter squadrons to Cigli AB, Turkey, and to Aviano AB, Italy, in 1966. In 1970, Zaragoza Air Base, Spain, was added to these rotations, which continued until summer 1970.

The 20th Tactical Fighter Wing was relocated from Wethersfield to RAF Upper Heyford, replacing the 66th Tactical Reconnaissance Wing, which had relocated from Laon Air Base, France, to the UK in June 1970. Once the 20th TFW arrived at RAF Upper Hayford, all aircraft were replaced by the General Dynamics F-111E Fighter-Bomber.

The British and American governments agreed to install a rapid engineering deployable heavy operational squadron, also known as RED HORSE, unit in the UK. The unit was established to manage the organisational issues for units at RAF Wethersfield and all other US air bases in Europe. The 819th and 2166th Communications Squadron were the main units left at RAF Wethersfield until the USAF returned the base back to the British in 1990. The 819th was deactivated in the early months of 1990, while the 2166th remained active until 1992. The base was handed back to the RAF at a ceremony on 3 July 1990, and Wethersfield was once again placed under care and maintenance status. The USAF retained a military family housing area at Wethersfield, in which American personnel assigned to RAF Molesworth, RAF Alconbury, resided.

In 1991, the Chief Constable of the Ministry of Defence Police (MDP) took responsibility for RAF Wethersfield. The Ministry of Defence Police Training School and Firearms Training was there as well as being home for the MOD Guard Service Training School (MSG). The move in 1994 of the MDP training and headquarters, along with the MGS Training Wing, gave the force its first combined HQ and Training Centre.

RAF Upper Heyford was a RAF air base, located north-west of Bicester in Oxfordshire, England. It opened in 1916 and was used by the RFC, but flying did not commence until 1918, when the RAF was formed. Upper Heyford continued to be a training facility until the 1950s and the outbreak of the Cold War. During the Cold War, Upper Heyford served as a base for the USAF Strategic Air Command (SAC) strategic bombers and later USAFE tactical reconnaissance, fighter, and fighter-bomber aircraft in the UK.

Due to ongoing worldwide threats, the SAC believed it necessary to build a formidable force of American air bases in England. It was decided to transform four airfields around Oxfordshire to serve as regular bases for the USAFE. Upper Heyford was one of those selected, the others being RAF Brize Norton, RAF Fairford, and RAF Greenham Common. Work started in June 1950 to extend the runway and add new hardstands for heavy bombers. A secure weapons storage facility was also built. In July 1950, the first group of USAF personnel arrived at Upper Heyford. It was designated the 7509th Air Base Squadron, and the 7509th would act as the host organisation to support the Temporary Duty Assignment (TDY) aircraft and personnel detached from their home bases in the US.

In May 1951, Upper Heyford was formally handed over to the USAF 3rd Air Force. Visiting TDY rotational units at Upper Heyford consisted of 93rd Bomb Wing, 97th Air Refuelling Squadron, 509th Air Refuelling Squadron, 301st Bomb Wing, 8th Air Sea Rescue Squadron, 2nd Bomb Wing, 5th Bomb Wing Detachment, and the 22nd Bomb Wing. Also, in May 1951, the 7509th Air Base Squadron was redesignated as the 7509th Air Base Group, but at the beginning of 1952, the 7509th became the 3918th Air Base Group. December 1951 saw the first arrival of aircraft to Upper Heyford, which were the B-50Ds from the 328th Bomb Squadron.

By the summer of 1952, Upper Heyford was ready to handle a full complement of forty-five aircraft, and when the 2nd Bombardment Wing arrived, it deployed all three of its bombardment squadrons. The SAC squadrons and wings continued to be deployed to Upper Hayford throughout the 1950s and '60s. In 1958, the unit was redesignated the 3918th Combat Support Group, with the occasional visits by the massive B-52 bomber beginning at the end of 1960 and becoming more and more frequent over the next five years.

In the summer of 1952, a detachment of top secret U-2 strategic reconnaissance aircraft commenced operations from Upper Heyford. This was to determine the characteristics of latest Soviet weapons.

In February 1964, the unit was redesignated the 3918th Strategic Wing, and it was decided that regular detachments of SAC bomber aircraft to England would cease altogether, and both Fairford and Greenham Common were closed. Upper Heyford 'Reflex Alert' continued until 1965 and the very last B-47 detachment was stood down at RAF Brize Norton in March 1965.

In March 1966, French President Charles de Gaulle announced that France would withdraw from NATO's integrated military structure. The US was informed that it must remove its military forces from France by April 1967.

RAF Upper Heyford was now to serve as the new and urgently needed base for the RF-101s of the 66th Tactical Reconnaissance Wing (TRW), which had been stationed at Laon-Couvron Air Base, France. After rapid preparations had been made, the transfer of this unit was completed by September 1966. During 1968, it was announced that the 66th TRW was to convert to the RF-4C Phantom, and in March 1969, the first two Phantoms flew into Upper Heyford. The RF-101Cs were assigned to the 18th TRS and were limited to the daylight role, while the RF-4Cs were assigned to the 17th TRS and were capable of an all-weather day and night operation.

Unfortunately, 66th TRW was deactivated in 1970 and the RF-4C Phantoms were moved to their new home with the 86th TFW at Zweibrücken in Germany while the RF-101s returned to the States and joined the 363rd at Shaw Air Force Base, South Carolina.

In the summer of 1970, the headquarters of the 20th Tactical Fighter Wing (TFW) relocated from RAF Wethersfield to RAF Upper Hayford, and with its arrival saw the 20th TFW begin converting to a new aircraft. September 1970 saw the arrival of two General Dynamics

F-4 Phantom arriving at RAF Alconbury air show back in the 1980s.

F-111Es at RAF Upper Hayford with all being delivered by February 1971, and the 20th TFW being declared operationally ready later that same year.

In 1983, another squadron arrived at the base, making four squadrons at RAF Upper Hayford; with this came the activation of 42nd Electronic Combat Squadron (ECS). The 20th TFW deployed on numerous NATO and US operations from 1972 to 1993, which included Operations Ocean Safari, Dawn Patrol, and Highwood to name a few.

In April 1986, over twenty-five F-111s took off from RAF Upper Heyford as part of the attacking force against the raid on Libya. They were used as an airborne reserve for the F-111Fs of the 48th TFW, from RAF Lakenheath. Three EF-111s were air spares and returned to RAF Upper Hayford, but on their return, they formed up with the 48th's F-111Fs and provided electronic defence during the attack on Tripoli.

In 1991, four squadrons from the 42nd ECS were reassigned to the 20th TFW. In January, aircraft from the 20th TFW launched combat missions during Operation Desert Storm from Turkey and Saudi Arabia and this continued until the ceasefire was declared. Many targets were attacked during the Iraqi War, including power plants, petroleum refineries, airfields, nuclear-biological-chemical processing and storage facilities, and electronics sites throughout northern Iraq. The wing had deployed 450 military personnel, flown 1,800 combat sorties without any loss, and dropped 4,700 tons of ordnance.

The occupancy of the 20th TFW was deemed no longer required after the Cold War, and the presence of the USAF at RAF Upper Hayford was gradually phased down. The 20th TFW, 55th, 77th, and 79th Tactical Fighter Squadrons were redesignated the 20th Fighter Wing (FW) in September 1991.

In 1993, flying activity was closed, and in early 1994, the 20th FW was deactivated at RAF Heyford and transferred to Shaw Air Force Base, South Carolina. Later that year, the base was returned to the Ministry of Defence, and today, the area is mainly residential buildings, rented accommodation, and shops and services that service the community.

The General Dynamics F-111 Aardvark at the Royal International Tattoo, 1991.

RAF Little Rissington is a RAF airfield and former RAF station in Gloucestershire, England. It was once home to the RAF's Central Flying School and the Red Arrows. It was opened in 1938 and finally closed in 1994. The main technical site was sold in 1996, but the aerodrome is maintained by the Ministry of Defence and remains active under the operational control of No. 2 Flying Training School at RAF College Cranwell.

It is now home to 637th Volunteer Gliding Squadron—the primary military unit—which provides elementary flying training for Combined Cadet Force and Air Training Corps Cadets. RAF Little Rissington Airfield is also used by the forces as a relief landing ground, training area, and parachute dropping area. The estate has also been used as a film set, including the filming of *The Avengers*, part of the ice chase in *Die Another Day*, and *Thunderbirds*.

During the Second World War, the flying training school and No. 8 Maintenance Unit occupied RAF Rissington and as the war events increased so did the number of aircraft being stored at the station. Due to security concerns, the level of security protection stepped up during the war, including the station's own fighter force of several Spitfires. Subsequently, various satellite airfields were used to spread out the increased number of aircraft.

In 1946, the RAF's Central Flying School moved to Little Rissington. During the same period, the station became home to the RAF's aerobatics teams, which included the Red Pelicans and later the Red Arrows.

With the arrival of the USAFE during the 1980s, for over ten years, Little Rissington became the largest military contingency hospital in Europe. The airfield regularly saw visiting C-130 Hercules and C5 Galaxies. During the Gulf War, Little Rissington was held on its highest readiness state of alert for several decades as it prepared for military casualties. The USAF left Little Rissington in 1993 and it was handed back to the RAF.

It is a local belief that a nuclear-proof underground hospital built by the USAF lies underneath RAF Little Rissington. The rumours, which have never been verified, were that, at the time, tunnels were dug during the Second World War, which housed the hospital.

RAF Manston was an RAF station located in north-east of Kent and was open from 1916 until 1996. The airfield was then reopened as a commercial airport but also continued being used by the military by the Defence Fire Training and Development Centre (DFTDC), continuing from a long-standing training facility for RAF firefighters at Manston.

With the outbreak of the Second World War came the arrival No. 3 Squadron at RAF Manston, equipped with Hawker Hurricanes and put under the command of No. 11 Group Fighter Command. During the Battle of Britain, RAF Manston was heavily bombed, with diary entries recording a steady stream of damage to aircraft and buildings during the war. RAF Manston was used during the testing period for Barnes Wallis and the famous 'Dam Buster' raid's Bouncing Bomb.

RAF Manston was developed as emergency landing grounds for bomber crews. The airfield was intended for use by returning bombers suffering from low fuel or suspected damage to the airframe. The location of RAF Manston on a hilltop was wisely chosen as it was usually fog free and had no approach obstructions. Being the closest frontline airfield, it saw most of badly damaged aircraft arrive. Most aircraft that had suffered from ground fire, collisions, or air attack, but retained a degree of airworthiness, landed at RAF Manston. The airfield became something of a 'boneyard' for heavy bombers and other aircraft landing, and it sometimes provided spare parts for other Allied aircraft in need of repair.

The 1950s and beginning of the Cold War saw the USAF use RAF Manston as a SAC base for its fighter and fighter-bomber units. The USAF SAC's backbone bombers were the Convair B-36 and Boeing B-47 Stratojet. By the 1950s, RAF Manston had only begun to recover from the devastating attacks of the Second World War. The control tower at Manston overlooked the hilltop runway, which was an astonishing 750 feet wide and 9,000 feet long. The 7th Air Division (AD) expanded Manston by building concrete bunkers, which were suitable for nuclear weapons and upgraded the support facilities for long-term use.

During the summer of 1953, the 7th AD began a series of deployments of B-47 and B-36 wings from the US to the UK. These deployments generally involved a huge amount of aircraft that included about twenty KC-97 Stratofreighters, which were maintained at the other English bases for around ninety days. At the end of the Temporary Duty (TDY), another SAC wing would take over command, and these deployments were continuous until 1955, when deployments moved to RAF Fairford and Manston was turned over to the USAFE.

In 1951, 12th Fighter-Escort Wing (FEW) deployed to RAF Manston to provide fighter escort for its bombardment wings, only remaining at Manston for the winter of that year when it was replaced by the 123rd Fighter-Bomber Wing. The 12th was transferred to Japan for combat duty during the Korean War.

The 123rd began to utilise the F-84E 'Thunderjets' that were left behind by the 12th FEW and continued the same mission of fighter escort of SAC's bombers. The year 1952 saw a change of wings, with the 406th Fighter-Bomber Wing being activated in place of the 123rd FEW at Manston.

The year 1954 saw a change in missions for the 406th from a fighter bomber to a fighter interceptor squadron; subsequently, a change of equipment saw the F-86D Sabre interceptor begin to arrive to RAF Manston and the F-86Fs were transferred to other USAFE squadrons and NATO countries.

In 1958, the 406th was deactivated, with its three air defence squadrons being assigned to continental Europe under the 86th Air Division at Ramstein Air Base, West Germany. The F-86Ds were withdrawn from Europe in 1961, and once the transfer of the USAFE interceptors at Manston was completed, the base was returned to RAF control. With the USAF withdrawal from RAF Manston, the airfield became a joint civil and military base with occasional package tour and cargo flights, while continuing its role as an RAF base.

RAF Manston was designated as one of the UK's Military Emergency Diversion Airfields (MEDA) for emergency military and civilian landings, and for several years, the base operated as a Master Diversion Airfield, open twenty-four hours every day. RAF Manston was a unique airfield in the UK, having a 'foam carpet' crash-landing system, where two tractors would pull tankers laying a metre-thick layer of foam over a strip of runway for aircraft with landing problems.

RAF Manston became home to the RAF's 22 Squadron from 1961, operating the Westland Whirlwind aircraft. They continued search and rescue support, with 72 Squadron taking over operations with the Westland Wessex in 1974, with the flight being transferred back to 22 Squadron in 1976. RAF Manston operated a Sea King from 202 Squadron search and rescue from 1988 until 1994, when SAR activity at the base was stopped and SAR cover for the Channel relocated to RAF Wattisham.

For many years, commercial airliners could be seen arriving and departing Manston. Invicta Airways and Air Ferry transported thousands of holiday passengers starting their journeys from Manston. From 1989, Manston became known as Kent International Airport, and a new terminal was officially opened that year by the Duchess of York.

It was decided to close the RAF Manston base in 1999. The 'airside' portion of the base was signed over to the commercial operator of Kent International Airport, but the MOD decided to keep the central fire training school (CTE) facility open. Unfortunately, Manston was closed in May 2014 and local rumours are that Manston will become a housing development.

RAF Barford St John is a RAF station just north of the village of Barford St John, Oxfordshire. It is now a non-flying operational facility, operated by the USAF as a communications centre with many large communications aerials. It is also a satellite facility of RAF Croughton. RAF Barford St John was opened in July 1940 as a training camp for the RAF Flying Training Command. Boasting three runways, the base was primarily used by the No. 15 FTS from nearby RAF Kidlington. The station was closed in late 1941 and was re-established as an RAF Bomber Command base and reopened in 1942.

The airfield gained paved runways and was equipped for night operations and was a satellite for RAF Upper Heyford. Bomber Command and No. 16 OTU was stationed there, flying the Vickers Wellingtons until December 1944.

No. 1655 de Havilland Mosquito Training Unit replaced the Wellingtons and the unit was renamed No. 16 OTU in 1945 when it moved to RAF Cottesmore. After the war, the airfield was closed in 1946 and placed into care and maintenance. In 1951, the USAF arrived at the base and opened a communications centre on the airfield, under the command of 2130th Communications Squadron at RAF Croughton.

Since the 1980s, the masts have been slowly dismantled and the eastern building compound was demolished in the late 1990s. Technically, the site is still in use but there is little activity. The Second World War concrete airstrips and taxi-ways are still intact, but, unfortunately, the Second World War buildings have been removed, being replaced by modern buildings on the airfield, secured and guarded with fencing and other security devices.

RAF Bentwaters is a former RAF station located near Woodbridge, Suffolk. Its name was taken from two cottages ('Bentwaters Cottages') that had stood on the site of the main runway during its construction in 1943. RAF Bentwaters was used by both the RAF during the Second World War and the USAF during the Cold War.

It was the primary home to the 81st Fighter Wing of the USAF between 1951 and 1993. The 81st Fighter Wing also operated from RAF Woodbridge during the same period.

Construction begun on a RAF station called RAF Butley in 1942, which was to be used by RAF Bomber Command, but in early 1943, the station was renamed RAF Bentwaters. The station was officially opened for operations in December 1944 when RAF Fighter Command from 11th Group took occupancy of RAF Bentwaters.

During the period of 1944 to 1945, over ten different squadrons were based at RAF Bentwaters, all flying various aircraft. In addition to its RAF use, USAAF fighters flew escort missions for RAF Bomber Command from Bentwaters, which begun in May 1945. After the Second World War, the RAF retained command of Bentwaters before closing the airfield in August 1949 and putting the site in to care and maintenance status.

In 1951, the command of RAF Bentwaters was transferred to the USAF and the USAFE. RAF Bentwaters became the headquarters of the USAFE in September 1951, playing a vital role in the defence of Western Europe during the Cold War era. In the early months of 1951, the USAF 7506th Air Support Group was assigned to RAF Bentwaters with the mission to expand and upgrade the facility. This took over twelve months to achieve.

The 81st Fighter Interceptor Wing (FIW) arrived at its new home in 1951 and it remained at RAF Bentwaters for over forty years. The 81st FIW housed three fighter interceptor squadrons, all flying the North American F-86A Sabre. The 81st FIW's operational squadrons were 91st Fighter-Interceptor (blue); 92nd Fighter-Interceptor (yellow); and 116th Fighter-Interceptor (red).

However, the 116th FIS was an activated Washington Air National Guard squadron that remained at Shepherds Grove, California, while assigned to the 81st. In November 1952, when the squadron was returned to state control, it was redesignated the 78th Fighter-Interceptor squadron. The 81st FIW's mission was to provide air defence to the UK and perform missions jointly with the RAF, working with No. 11 Group RAF Fighter Command. The wing was the first unit of F-86s assigned to NATO and the first foreign air unit to participate in the peacetime air defence of the UK.

Missions changed for the 81st Fighter Interceptors to Fighter-Bomber operations in 1954 with the wing supporting operations for the USAFE and NATO. With this change saw the unit redesignated as the 81st Fighter-Bomber Wing, being re-equipped with the Republic F-84 'Thunderstreak'.

In 1958, Bentwaters was operated as 'twin base' with RAF Woodbridge and as a single unit under the newly designated 81st Tactical Fighter Wing (TFW). The 91st and 92nd Squadrons from the 78th Tactical Fighter Squadron were transferred from RAF Shepherds Grove to Bentwaters and the 78th Tactical Fighter Squadron operated from RAF Woodbridge.

In late 1958, the 81st TFW was re-equipped with the McDonnell F-101A/C 'Voodoo'. The F-101 was configured as a fighter-bomber, intended to carry a single nuclear weapon for use against battlefield targets such as airfields. The F-101s primary mission was delivering nuclear weapons at extremely low altitudes. Pilots were trained for one-way missions into Soviet territory to increase effective range at some cost in negating pilot recovery.

In the winter of 1965, the 81st received the McDonnell F-4 Phantom IIs to replace the F-101s. Initially receiving the F-4C, this was later upgraded to the more capable F-4D during late 1972

The Fairchild Republic A-10 Thunderbolt II arriving at Royal International Air Tattoo, 1991.

to 1973. The tail codes for the 81st TFW F-4s were 78th Tactical Fighter—WR (red tail stripe); 91st Tactical Fighter—WS (blue tail stripe); 92d Tactical Fighter—WT (yellow tail stripe).

In 1979, the 81st TFW began its conversion to the Fairchild Republic A-10 Thunderbolt II. The A-10 was a single-seat, twin-engined jet aircraft designed to provide close air support (CAS) of ground forces by attacking tanks, armoured vehicles, and other ground targets. With the conversion to the A-10, the wing's mission changed to close air support and battlefield air interdiction in support of NATO ground forces.

The 81st TFW expanded to six A-10 squadrons, which were shared between RAF Woodbridge and RAF Bentwaters: 78th Tactical Fighter (Woodbridge); 91st Tactical Fighter (Woodbridge); 92nd Tactical Fighter (Bentwaters); 509th Tactical Fighter (Bentwaters); 510th Tactical Fighter (Bentwaters); and 511th Tactical Fighter (Bentwaters). The 509th and 511th were transferred to RAF Alconbury in 1988 when the 10th Tactical Reconnaissance Wing became a Tactical Fighter Wing.

With the transfer of the A-10s to RAF Alconbury, the 527th Aggressor Squadron was transferred to RAF Bentwaters. The 527th was to provide air combat training for NATO pilots, using Soviet tactics. The squadron flew the Northrop F-5 Freedom Fighter's at RAF Alconbury; however, after several years of intense flying, the fleet of aggressor F-5Es was to be replaced and the entire squadron was re-equipped with General Dynamics F-16C Fighting Falcons when the squadron was assigned to RAF Bentwaters.

The 527th AS flew its last F-5E sortie from RAF Alconbury in June 1988, and in July 1988, the squadron was re-equipped to F-16Cs at Bentwaters. With the collapse of the Soviet Union, the decision was made to terminate the entire USAF aggressor program and the 527th AS were deactivated in September 1990.

With the end of the Cold War, the USAF presence at RAF Bentwaters was slowly phased down and it was announced that the station would be closed and the 81st TFW would be deactivated. The 510th Tactical Fighter Squadron was transferred to the 52nd Fighter Wing at Spangdahlem Air Base, Germany, in 1993, but was deactivated the following year. The 510th was again reactivated when it moved again that same year to Aviano Air Base, Italy, re-quipped with the F-16C/D.

The 92nd Tactical Fighter Squadron was also deactivated in 1993 and the aircraft were sent back to the States, joining various Air National Guard squadrons. With the deactivation of all the wings, the USAF returned control of RAF Bentwaters to the Ministry of Defence in 1993.

In 2007, the Bentwaters Cold War Museum opened its doors to the public. It is located in the former USAF hardened command post and is believed to be the only such building open to the public in the UK. The main 'war operations room' and 'battle cabin' have been restored to their original condition. The museum continues opening its doors on special occasions and events for the public and is run by volunteers from the Bentwaters Aviation Society.

RAF Blenheim Crescent is a non-flying RAF station located in Ruislip in Greater London and is presently used by the US military for administrative purposes.

The 422nd Air Base Group based at RAF Croughton is predominately in command of Blenheim Crescent, with a collection of others operating support agencies. The US Navy held the lease until 2007, when control was handed over the USAF in the winter. The facility at present has undergone renovation and is planned for the US Navy, Army, and Air Force to be stationed here.

RAF Burtonwood is a former RAF station 2 miles north-west of Warrington, Cheshire. RAF Burtonwood airfield was used in January 1940 as a servicing and storage centre for modifying British aircraft and was operated by No. 37 Maintenance Unit until 1942.

In June 1942, RAF Burtonwood was transferred to the USAAF and became the command centre for the US Eighth, Ninth, Twelfth, and Fifteenth Air Forces' aircraft. RAF Burtonwood was boasted to be the largest airfield in Europe during the Second World War, with the most USAAF personnel and aircraft maintenance facilities. By the end of the Second World War, over 18,000 servicemen were stationed at RAF Burtonwood.

RAF Burtonwood was returned to the RAF in 1946 and became an equipment depot; however, later that year, six B-29 Superfortress bombers from the USAAF SAC 43rd Bombardment Group

Boeing B-29 Superfortress B-29A 44-61748 is now housed at the Duxford Air Museum.

was sent to Burtonwood as a 'training deployment'. In May 1947, additional B-29s were sent to RAF Burtonwood to keep up the presence of a training programme.

These deployments were to have a strategic air force permanently stationed in Europe. In November 1953, the arrival of the 53rd Weather Reconnaissance Squadron flying the WB-29 occurred, and then the WB-500 Super Fortress. The squadron's mission was to collect weather data for the air force. The squadron was transferred to RAF Alconbury in 1959.

The Military Air Transport Service (MATS) was also based at RAF Burtonwood, which housed a cargo and passenger facility until 1958 when the operation was moved to RAF Mildenhall, where transatlantic transport flights through the base were operated by Douglas C-54, Douglas C-118, Douglas C-124, Douglas C-133 Cargomaster, and Lockheed C-130 aircraft. RAF Burtonwood's USAF operations ended in April 1959, and for several years, the facility fell into disuse. The USAF returned the station to the Ministry of Defence in 1965.

During 1970s and the 1980s, the area was used by the Territorial Army and Cadet units for training purposes. The site was also used by the MOD for civil contingency and emergency planning exercises, as well as EOD exercises for police, fire, and rescue training.

RAF Chelveston is a former RAF station located 5 miles east of Wellingborough, near the village of Chelveston in Northamptonshire. During the Second World War, the airfield was occupied by both the RAF and the USAAF. During the Cold War, RAF Chelveston housed some flying units, but its leading role was to be a station ready to receive USAF units in case of an emergency. In 1975, the airfield was sold by the Ministry of Defence to private land owners; however, part of the airfield was retained as a military housing area occupied by US service members assigned to RAF Alconbury and RAF Molesworth.

RAF Chelveston began construction in 1940 with grass runways and three hangers. The airfield opened in the summer of 1941 as a RAF station. The airfield was upgraded for the

Douglas C-47 42-100882 was delivered to the USAAF in 1943, seen here at the Duxford Air Show.

arrival of the US Eighth Air Force and was used as a heavy bomber airfield in 1942. The first USAAF unit to occupy Chelveston was the 60th Troop Carrier Group. The 60th consisted of the 10th, 11th, 12th, and 26th Squadrons, equipped with over fifty C-47 aircraft. The 60th arrived in early July 1942 but moved to RAF Aldermaston at the end of the month. After training in the UK, the unit moved on to the 12th Air Force for operations in the Mediterranean theatre.

In August 1942, the 301st Bombardment Group (BG) took up occupancy at the station. The 301st was assigned to the 1st Combat Wing at Brampton Grange. Its operational squadrons were the 32nd, 352nd, 353rd, 354th, and 415th Bomb Squadrons, each equipped with Boeing B-17F Flying Fortresses. With the unit's arrival, it was the second heavy USAAF bomber group to arrive in England. The wing flew its first operational mission in September 1942 to northern France. During its stay at RAF Chelveston, the unit attacked submarine pens, airfields, railroads, bridges, and other targets in continental Europe, but primarily in France.

The 301st BG was allocated to Operation Torch, and in November 1942, it departed for the Twelfth Air Force and relocated to Tafaraoui Airfield, Algeria.

In the winter of 1942, the airfield experienced the arrival of the 305th Bombardment Group, transferring from RAF Grafton Underwood. The 305th BG was assigned to the 40th Combat Wing at RAF Thurleigh. Its operational squadrons were the 364th, 365th, 366th, and 422d Bomb Squadrons, each initially equipped with B-17Fs, then upgraded to B-17Gs in 1944.

The 305th BG missions were to bomb naval yards in Germany, and in January 1943, the first bombers from the Eighth Air Force penetrated Germany. The group attacks continued throughout 1943 with several strategic targets, including submarine pens, docks, harbours, shipyards, motor works, and marshalling yards in France and Germany.

Boeing B-17G *Sally B* accepted by the USAAF as 44-85784 in 1945, showing the underside of the aircraft.

The 305th moved to St Trond Airfield, Belgium, in 1945, and RAF Chelveston was handed back to the RAF in the winter of 1945. The airfield was placed under care and maintenance immediately after the Second World War ended.

The US and UK government agreed to expand the US military presence in the UK due to the growing threat of the Soviet Union. As such, RAF Chelveston return to USAF control in 1952, and the USAF planned to use Chelveston as a base for forward deployment of the new Boeing B-47 Stratojet medium range nuclear bomber.

SAC wanted its nuclear bomber force of B-47s stationed at forward bases in Western Europe and North Africa. The borders of the Soviet Union were within the range of the B-47.

Starting in November 1955, RAF Chelveston was transferred to the SAC 7530th Air Base Squadron, 7th Air Division, as a standby facility. The base was reconstructed, and a completely new runway was constructed on the airfield to accommodate intercontinental bombers. Wartime buildings in disrepair were replaced with new facilities.

There was a major adjustment of USAF deployments in Western Europe following French President General de Gaulle's requirement for all foreign nuclear forces to leave France, and the B-47 deployments abruptly ended.

In 1959, the command of RAF Chelveston was transferred from SAC to USAFE. Control of the base was assigned to the 10th Tactical Reconnaissance Wing at RAF Alconbury, which had been moved from West Germany to England. The 10th TRW could not accommodate all four squadrons of the wing at Alconbury, so one was based at Chelveston.

The 19th TRS remained at RAF Chelveston for three years, and it was decided to forward deploy the squadron to the then-unused Toul-Rosières Air Base, France, in 1962. After the B-66s departed, the base was returned to reserve status. The station was maintained by a small RAF skeleton support staff.

Due to budget cuts in 1970, the MOD decided to close RAF Chelveston, and in 1977, the runways and hardstands were removed. Other parts of the base were sold to private agricultural interests, except for some housing units retained by the USAF for families of personnel assigned to RAF Molesworth.

In 2005, RAF Station Chelveston was sold by Defence Estates to a businessman, and in May 2007, a new memorial to the men that served at the RAF station during the Second World War was unveiled in their memory by some of their comrades.

RAF Chicksands is a former RAF station located 7.7 miles south-east of Bedford, Bedfordshire. The estate was bought by the Crown Commissioners in 1936. Gerald Bagshawe lived on the estate until it was requisitioned by the Royal Navy. The RAF took control of operations a few months later and established a signal intelligence collection (SIGINT). SIGINT was established on the site throughout the Second World War, intercepting German traffic and passing the resulting material to the Government Code and Cypher School at Bletchley Park.

In 1950, the site was subleased to the USAF, serving as the base of the 6940th Radio Squadron, responsible for continued communications and SIGINT operation through the Cold War. The RAF continued to act as a host unit for the resident USAF units, including over time the 6950th United States Air Force Security Squadron, later becoming the 6950th Electronic Security Group and the 7274th Air Base Group.

In 1962, a 1,443-foot antenna was constructed at RAF Chicksands to form part of the Iron Horse HF direction finding network. This antenna, dubbed the Elephant Cage, was dismantled in 1996 when the USAF withdrew from the site, handing it back to the British Armed Forces.

In 1997, the British Army Intelligence Corps assumed responsibility for the site, moving the Corps' HQ from Ashford, Kent, along with Intelligence Training. Since then, the site has been the home of the Defence Intelligence and Security Centre (DISC) and the HQ of the Intelligence Corps. DISC conducts training for personnel of all three arms of the British Armed Forces, members of the Civil Service, and others. Courses are delivered across the range of intelligence disciplines.

RAF Croughton is a RAF station that is currently a USAF communications station in Northamptonshire, England. The station is home to the 422nd Air Base Group and operates one of Europe's largest military switchboards and processes approximately a third of US military communications in Europe.

RAF Croughton houses the 422nd Air Base Group and its mission is to provide installation support, force protection, and worldwide communications across the entire span of operations. The group located in the UK supports NATO, US European Command, US Central Command, Air Force Special Operations Command, US Department of State operations, and MOD operations.

RAF Coughton was built in 1938 and was first known as Brackley Landing Ground, becoming RAF Brackley in 1941. The name changed again to RAF Croughton. The site consolidated of three farms with the three grass runways. In June 1940, the station became a satellite airfield for RAF Upper Heyford for No. 16 OTU to provide extra airfield space for night-flying training.

Operations were with the command of the newly formed No. 7 Group of RAF Bomber Command, and the training was for commonwealth pilots (Canadian, South African, Australian, and New Zealand) on the Handley Page Hampdens, Bristol Blenheims, and Wellington Bombers.

In September 1940, the MOD decided that a selection of stations would be designated as emergency airfields. These stations would give assistance to any operational aircraft returning damaged. RAF Croughton would remain open with its airfield landing lights illuminated irrespective of enemy activity in the area. This meant that RAF Croughton, along with other emergency airfields, would attract enemy night attacks. RAF Croughton was at considerable risk of attack due to the fact it was a training airfield. Both sides during the war considered it worthwhile to attack and disrupt training programs, being lightly defended, they made for easier targets.

RAF Croughton remained a satellite for RAF Upper Heyford, but the base became the new home for the gliders in July 1942. During this training, Hawker Hectors, American Harvards, and Miles Masters pulled Hotspur gliders, while DC-3s pulled the larger Horsa gliders.

In March 1943, the last class passed out of RAF Croughton and the glider school closed, but the airfield continued to be a training base and became a satellite airfield to RAF Kidlington in Oxford. A new unit was formed, No. 20 AFU, with pilot training on bomber or coastal command.

In April 1943, No. 1538 Beam Approach Training Flight was formed at RAF Croughton and added to the station's training mission. Both No. 1538 BATF and No. 20 AFU remained at RAF Croughton until October 1944, when the airfield returned to Flying Training Command. On this date, No. 1538 BATF disbanded and No. 20 AFU retired to RAF Kidlington. In November 1944, the No. 1 Glider Training School (GTS) arrived at RAF Croughton until

1945, when command was transferred to No. 21 Hotspur Glider Conversion Unit at RAF Brize Norton. Training continued at RAF Croughton until 1946, when all training was ceased, and No. 1 GTS left RAF Croughton.

In the winter of 1950, the USAF took command of the station with the 1969th Communication Squadron at South Ruislip and was formed as a detachment at RAF Croughton. This began RAF Croughton's new communications mission, and over the next several decades, the units stationed at RAF Croughton changed many times, but the mission remained communications. In 1955, this detachment became the 1230th Airways and Air Communications Service Squadron (AACS) as part of the establishment of the Air Force Communication Service as a separate major command—Air Force Communications Command. In 1961, No. 1230 AACS was redesignated to become No. 2130th Communications Squadron (CS). With the mission and unit growth, the squadron became No. 2130th Communications Group (CG).

In 1980, the formation of No. 2147th Communications Group at RAF Mildenhall occurred, as well as the realignment of many squadrons and operating locations. No. 2130 CG deactivated several communications missions at RAF Croughton with the responsibility for their other missions falling on No. 2168th Communications Squadron at RAF Upper Heyford. No. 2130th Communications Squadron was reactivated in July 1983 and assigned to manage the communications missions at RAF Croughton. The mission increased in December 1985 when the Giant Talk station at RAF Croughton began operations. In 1986, No. 2130 CS was again redesignated to become No. 2130th Communications Group.

The year 1993 saw several substantial changes at RAF Croughton. Most of these had to do with mission support. With the projected deactivation of the 20th Fighter Wing at RAF Upper Heyford and closure of that station, the 2130 CG was redesignated as the 630th Communications Squadron. This new squadron fell under the command of the 100th Communications Group (CG) at RAF Mildenhall.

Due to the closure of RAF Upper Hayford, RAF Croughton began to receive most of its mission support from the 100 CG as well as some from the 100th Regional Support Group (RSG). Both units were housed at RAF Mildenhall under the 100th Air Refuelling Wing (ARW).

In July 1994, the USAFE conducted another command-wide reorganisation. The reorganisation meant the deactivation of the 100 RSG, but in its place, the USAFE activated the 603rd Regional Support Group as an independent group directly under Third Air Force. This in turn forced the redesignation of the 630th CS to the 603rd Communications Squadron.

In 2004, the 420th Air Base Group took occupancy at RAF Fairford and reported directly to the Third Air Force. The 38th Combat Support Wing (CSW) was formed to provide support to three GSU Groups, the 38th CSW and 420th ABG were responsible for the operations of the 422nd ABS at RAF Croughton.

In 2005, the 422nd ABG became a unit of the 501st CSW, replacing the 38th CSW. The change created other squadrons, which included the 422nd Air Base Squadron, the 422nd Communications Squadron, the 422nd Civil Engineering Squadron, and the 422nd Security Forces Squadron.

RAF Feltwell is a RAF station located 10 miles west of Thetford in Norfolk, and it was used by the USAFE. Formerly a Second World War bomber station, the airfield is now used as a housing estate for USAF personnel stationed nearby at RAF Mildenhall as part of the 100th Air Refuelling Wing and RAF Lakenheath as part of the 48th Fighter Wing.

The airfield was constructed for the RAF in the late 1930s and it was home to several heavy bomber squadrons of the RAF during the Second World War. The Thor missiles were stationed at RAF Feltwell at the time of the Cuban Missile Crisis. In 1963, after the departure of the Thor missiles, the RAF's Officer Cadet Training Unit was based there.

Between 1989 and 2003, it also hosted the USAF's 5th Space Surveillance Squadron (5 SPSS) which was lower ranking to the 21st Operations Group (21 OG) and the 21st Space Wing (21 SW), both housed at Peterson AFB, Colorado, which reports to Headquarters Air Force Space Command (AFSPC), also at Peterson AFB, CO.

RAF Feltwell is now administered by Detachment 4 of the 18th Intelligence Squadron, which is a space control intelligence organisation of the USAF, located at Wright-Patterson Air Force Base, Ohio.

RAF Menwith Hill is a RAF station near Harrogate, North Yorkshire. RAF Menwith Hill provides communications and intelligence support services to the UK and the US. The station contains an extensive satellite ground station and is a communications intercept and missile warning site. It is the largest electronic monitoring station in the world.

RAF Menwith Hill was commanded by a RAF officer, supported by an RAF element, with support services also provided by the USAF, 421st Air Base Group, until 2014. The site acts as a ground station for several satellites operated by the US National Reconnaissance Office, on behalf of the US National Security Agency, with antennae contained in many highly distinctive white radomes; it is alleged to be an element of the ECHELON system.

In 1954, the land was acquired by the British War Office and leased to the US. The US Army Security Agency was established to monitor communications from the Soviet Union.

The National Security Agency (NSA) took responsibility for the US operation at RAF Menwith in 1966. The NSA automated system 'watch-list' scrutiny of intercepted messages that had not been enciphered.

RAF Menwith Hill's primary mission is to provide intelligence support for UK, US, and allied interests. The bases went through a multimillion-pound expansion called 'Project Phoenix', and it is one of the largest and most sophisticated high technology programs carried out anywhere in the UK. It houses 1,800 employees, with over 400 British and 1,200 American employees of the NSA.

During the 2009 G-20 London Summit, NSA intercept specialists based at RAF Menwith Hill attempted to target and decode the encrypted telephone calls of the Russian president.

For over a decade beginning 1984, several peace camps were established near RAF Menwith. Many protests were predominantly related to nuclear strategic missile defence. Other protests also occurred at US military locations in the UK, such as Greenham Common. Objections to the Strategic Defense Initiative led to demonstrators storming the perimeter fence. In April 2012, activists affiliated with the global Occupy Movement held a four-day campsite at RAF Menwith Hill.

RAF Molesworth is a RAF station located near Molesworth, Cambridgeshire. It is now a non-flying facility under the control of the USAF, with a history dating back to 1917. RAF Molesworth saw its first flying unit during the First World War, when No. 75 Squadron from the RFC arrived, extensively flying the B.E.2 (a reconnaissance aircraft) for combat over the Western Front. After the First World War, the airfield was abandoned with some farm owners taking over occupancy of some building on the site.

On the outbreak of the Second World War, the Air Ministry selected the site, and the building of RAF Molesworth began in 1940. The first unit to arrive at RAF Molesworth was the RAAF 460 Squadron, which arrived in November 1941 flying the Vickers Wellington IVs. No. 460 Squadron departed RAF Molesworth a year later when 159 Squadron (RAF Bomber Command) arrived; however, the unit did not remain long and moved to the Middle East in early 1942.

The first USAAF unit to arrive at RAF Molesworth was the 15th Bombardment Squadron, arriving in June 1942. The squadron flew the Douglas Boston III (A-20) light bomber. The 15th was originally part of the 27th Bombardment Group (Light), assigned to Fifth Air Force in the Philippine Islands. The group's aircraft did not arrive until December 1941, and due to the deteriorating situation in the Philippines after the Japanese invasion, they were diverted to Australia.

The 15th flew most of its missions from RAF Molesworth in Boston aircraft, and it did not receive USAAF Douglas A-20 Havocs until September 1942. The squadron was transferred to RAF Podington until the middle of September 1942, flying a few missions before being transferred to Twelfth Air Force for support of Allied landings in North Africa in October 1942.

When the 15th Bomb Squadron departed RAF Molesworth, it was reoccupied by the Boeing B-17 Flying Fortresses of the 358th Bombardment Group. This was the first of four squadrons that would join the 303rd Bombardment Group. The 303rd remained at RAF Molesworth until after V-E Day in 1945.

In November 1942, the 358th flew its first mission. Initially, missions targeted subjects such as aerodromes, railways, and submarine pens in France until 1943. The group would become one of the legendary units of the Eighth Air Force. In January 1943, the 303rd took part in the first penetration attacks in Germany by heavy bombers of Eighth Air Force striking U-boat yards, then attacking other targets such as ball-bearing plants, shipbuilding yards, and aircraft engine factories.

The 303rd flew its last mission on 25 April. During its combat tour, the group flew over 360 missions, comprising 10,270 sorties, 26,346 tons of bombs dropped, and 378 enemy aircraft shot down. The group also saw over 800 of its men killed in action with another 750 becoming prisoners of war. In May 1945, the 303rd Bomb Group left RAF Molesworth, being deactivated in Morocco in July 1945.

In July 1945, the command of RAF Molesworth was given back to the RAF, and the base was chosen to become a training airfield for their new jet aircraft. The first jet unit, 1335 Conversion Unit, arrived in July, flying the Gloster Meteor. It would be joined over the next year by several transient aircraft on deployments. A year later, the 1335 Conversion Unit moved from RAF Molesworth and the airfield was then placed under 'care and maintenance'.

During increased tension and the outbreak of the Cold War, the USAF began looking to expand in Western Europe. RAF Molesworth was chosen in 1951 to become home to the 582d Air Resupply Group. The station underwent expansion with main runway extensions and modern facilities.

In September 1953, the Korean peace treaty was signed, which ended active conflict on the Korean peninsula, and the USAF air supply wings were reduced. Although the unit was identified as an air resupply group, the unit's name was deliberately misleading, as the mission of the 582nd was support of special operations over Soviet occupied territory.

The 582nd was assigned directly to Third Air Force and provided the bulk of its air support to the Army 10th Special Forces Group. During the next two years, the 582nd worked closely with the 10th Group, providing airdrop, resupply, and air-land support with its assigned B-29 and C-119 aircraft.

In October 1956, the 582nd was replaced by the 42nd Troop Carrier Squadron. The 42nd flew HU-16s, C-47s, C-119s, and C-54 Douglas Sky master cargo transport aircraft from RAF Molesworth until May 1957, when it moved to RAF Alconbury. However, the squadron was deactivated in December 1957.

With the departure of the 42nd Troop Carrier Squadron, RAF Molesworth was put into a standby status again, with occasional aircraft using the airfield. In April 1959, with the closing of the active runway at RAF Burtonwood, the WB-50Ds of the 53rd Weather Reconnaissance Squadron used the runway at RAF Molesworth, although the unit was officially assigned to RAF Alconbury. The squadron was reassigned to RAF Mildenhall in August 1959. RAF Molesworth personnel was reduced to a small team of maintenance staff, and for the next twenty years, the facility was in reserve status with upkeep provided by RAF Alconbury.

In 1973, RAF Molesworth as an airfield was officially deactivated and the only remaining structures on the station were a small arm firing range, used by USAF personnel for target practice, and a compound for surplus and obsolete USAF vehicles manned by civilian USAF employees. The fire station housed one engine and a small crew for fire protection; these firefighters were USAF personnel from RAF Alconbury who were assigned to RAF Molesworth on a rotating basis. Two other structures also remained—the canteen and the chapel.

In June 1980, RAF Molesworth was selected as one of two British bases for the USAF's mobile nuclear armed Ground Launched Cruise Missiles, or GLCMs. Most GLCMs were deployed at RAF Greenham Common, the other UK base.

In December 1986, the USAFE 303rd Tactical Missile Wing was reactivated at Molesworth with BGM-109G Gryphon GLCMs. As RAF Molesworth no longer had a runway, the missiles were flown into RAF Alconbury before being transferred by road to RAF Molesworth. In 1987, the US and the Soviet Union signed the Intermediate-Range Nuclear Forces Treaty, which led to the removal of all nuclear missiles from RAF Molesworth by October 1988. The 303rd TMW were deactivated in January 1989.

In January 1990, the RAF announced construction would begin later that year to house the US European Command's new intelligence analysis centre. The Joint Analysis Centre (JAC) provided intelligence support for US and NATO missions in the Middle East and the Balkans while providing global assistance in the War on Terror.

RAF Molesworth is home to three Major Command (MAJCOM) branch sites, which include the US European Command (USEUCOM) Joint Intelligence Operations Centre Europe Analytic Centre (JAC), United States Africa Command (USAFRICOM) Intelligence and Knowledge Directorate-Molesworth (J2-M), North Atlantic Treaty Organization (NATO) Intelligence Fusion Centre (IFC) and is maintained by the USAF's 423rd Air Base Group (423 ABG), located at RAF Alconbury.

RAF Molesworth employs over 750 personnel, including US, British, and foreign military, as well as US and British civilians. RAF Molesworth relied solely upon the 423rd ABG for all non-JAC related support functions like postal services, banking, and telecommunications connectivity. In January 2015, it was announced that the USAF would be withdrawn from RAF Molesworth, along with RAF Mildenhall and RAF Alconbury.

RAF Greenham Common is a former RAF station located to the south-east of Newbury, Berkshire. Opening in 1942, the base was used by both the RAF and USAAF during the Second World War and the USAF during the Cold War.

In late 1943, Greenham Common airfield was under the command of the USAAF Ninth Air Force, with the first arrival being the 51st Troop Carrier Wing (TCW) HQ, arriving in September 1942. The 51st TCW controlled the three troop carrier groups (TCG) at RAF Keevil (62nd TCG), RAF Aldermaston (60th TCG), and RAF Ramsbury (64th TCG) as part of Twelfth Air Force. The TCGs began to arrive and was deployed to RAF Greenham Common. The Ninth Air Force used RAF Greenham Common airfield for fighter groups arriving to the UK. Arriving in November 1943 was the 354th Fighter Group (FG) from the States, which flew the North American P-51 Mustang. This was short lived, and the group moved to RAF Boxted a few days later.

In January 1944, the 368th Fighter Group arrived again from the States, this time flying the P-47 Thunderbolt. The group consisted of three squadrons in total, but again, its stay was short lived, moving the following day to RAF Chilbolton.

RAF Greenham Common was a very fast-paced airfield, and as the 368th FG moved out, the 438th Troop Carrier group flew into RAF Greenham Common, flying the Douglas C-47 Skytrain. The 438th TCG consisted of four squadrons.

The 316th TCG used Greenham Common as a forward base. With the end of the Second World War came the closure of RAF Greenham Common, which remained inactive for the next five years. In April 1951, RAF Greenham Common was made available to the USAF by the British MOD as a SAC base with joint operations with RAF units. Massive reconstruction work began with basically the entire wartime airfield demolished and replaced with a new runway that had parallel taxiways north and south of it, leading to extensive hard standings. Construction work was finally completed, and the base declared operational in September 1953.

RAF Greenham Common was taken over by the SAC's 7th Air Division, with the 3909th Combat Support Group as its administrative unit on the base.

The 303rd Bombardment Wing was deployed in March 1954, flying the B-47 Stratojets. The 303rd BW stayed at RAF Greenham Common for just over a month, returning on 28 April 1954. This was the first of several short-term temporary duty deployments from home bases in the US that continued intermittently over the next ten years.

In 1958, the runways and dispersals were upgraded and strengthened for the Boeing B-52 Stratofortress bomber, and in 1960, the B-52s made training visit to RAF Greenham Common. The SAC departed Greenham Common in June 1964, much to the relief of the residents, and for several years, it was used for USAF storage and as a relief base.

The closure of US bases in France forced the reopening of RAF Greenham Common for air transport operations. In late 1967, Greenham Common was used for NATO Exercise Reforger, again because of the withdrawal of France from the NATO integrated military alliance.

In November 1968, the control of RAF Greenham Common was transferred from SAC to the USAFE, with the 7551st Combat Support Group having administrative control of the base. However, the base was primarily utilised as a US Military Postal Mail sorting facility, with aircraft flying mail in from the US.

In 1977, the USAF announced plans to reactivate the base to house the KC-135 aerial refuelling tanker aircraft. This was due to a lack of capacity at their main base, RAF

Douglas C-47A Dakota 2100884, seen here doing a flypast at the Duxford Air Show.

Boeing B-52H 60-042 *The Red Gremlin II* from 93rd Bomb Squadron, Barksdale Air Force Base, arriving at RAF Fairford, 2012.

Mildenhall. This led to widespread local opposition, and in 1978, the British Defence Secretary vetoed the plan.

In June 1980, it was announced that RAF Greenham Common was to become the first site for cruise missiles, with ninety-six based at RAF Greenham Common and sixty-four at RAF Molesworth. With the announcement, a massive new construction was undertaken as the GLCM Alert and Maintenance Area (GAMA) site was built. GAMA was a maximum security QRA (Quick Reaction Alert) area with six large shelters in which fully operational cruise missiles were stored. The shelters were specially designed and constructed to protect against nuclear and conventional strikes. They were about 10 metres high, with a reinforced 2-metre-thick concrete ceiling. The shelters were completely covered with tons of clay. Each shelter was equipped with three hydraulic nuclear blast-proof doors at. They were designed to withstand the blast of an air-bursting nuclear explosion above the base or a direct hit from a 2,500-lb conventional bomb.

In 1983, the 501st Tactical Missiles Wing was the first squadron to receive its weapons, which were transported into the base by Lockheed C-5 Galaxys. The US and the Soviet Union signed the Intermediate-Range Nuclear Forces Treaty in 1987, which led to the removal of all nuclear missiles from the base. RAF Greenham Common removed the last GLCMs in March 1991, and the 501st TMW were deactivated on June 1991.

On 11 September 1992, the USAF returned RAF Greenham Common to the MOD. In February 1993, the Greenham Common air base was declared surplus to requirements by the Secretary of State for Defence and the facility was closed and put up for sale.

RAF Welford is an active RAF station in Berkshire. RAF Welford was opened in 1943 and was used during the Second World War by both the RAF and USAAF, predominately as a transport airfield. In 1946, the airfield was closed and placed in reserve status. Due to the Cold War, the station was reopened in 1955 as a munitions depot by the USAF.

In the present day, RAF Welford is one of the largest ammunition compounds for the USAF in Western Europe for heavy munitions. In October 1943, the airfield was allocated to Ninth Air Force, IX Troop Carrier Command (TCC), while under USAAF control. In November 1943, 315th Troop Carrier Group (TCG) arrived at Welford from RAF Aldermaston, flying C-47s and C-53s. The 315th TCG were part of the 52nd Troop Carrier Wing. In February 1944, the group was transferred to RAF Spanhoe.

As part of the IX Troop Carrier Command's desire to have their C-47 groups commence training with paratroops of the 101st Airborne Division deployed in the Salisbury Plain area, the squadrons of the 435th Troop Carrier Group arrived at Welford on January 1944 from RAF Langar, flying C-47s and C-53s.

The 435th Tactical Combat Wing (TCW) was assigned to the 53rd Troop Carrier Wing. In early February 1945, the group moved to an advanced landing ground at Breigny France (A-48).

In 2009, USAF staffing at Welford was reduced as part of USAF-wide budgetary adjustments. RAF Welford is now under the command of the 420 Munitions Squadron and comes under the command of the 501st Combat Support Wing, with HQ at RAF Alconbury.

RAF Alconbury is an active RAF station located in Huntingdon, Cambridgeshire. The station is currently under the control of the USAF but is a non-flying facility. It is one of three stations in operation in the Cambridgeshire area currently used by USAFE-AFAFRICA.

Lockheed-Martin C-5M 70033 from Westover Air Force Base, arriving at a grey overcast RAF Fairford, 2012.

Douglas C-47A Skytrain 330652/N345AB showing a great topside angle of the aircraft on a flypast to the Shuttleworth Air Show.

RAF Alconbury and RAF Molesworth are the last Second World War-era Eighth Air Force bases in England that are still actively in use and controlled by the USAF. In January 2015, it was announced that the USAF would be withdrawn from RAF Alconbury, along with RAF Mildenhall and RAF Molesworth, but since the newly elected President Donald Trump, the bases' futures are uncertain.

RAF Alconbury hosts the 423rd Air Base Group (423rd ABG), which supplies units for RAF Alconbury and RAF Molesworth, and provides services to the 426th Air Base Squadron at Sola Air Station, Stavanger, Norway.

The group houses six squadrons, which include security forces and civil engineers, medical services, and accommodation support units. The primary mission is to support the US European Command centre and National Geospatial-Intelligence Agency at RAF Molesworth. The medical and dental clinic at RAF Alconbury serves rapid medical needs to active duty personnel and their families plus retired military persons who live in the area.

RAF Alconbury is also the home of the 501st Combat Support Wing (501 CSW). The 501 CSW ensures UK-based air base groups are resourced, sustained, trained, and equipped to provide mission support that enables US and NATO forces to conduct full spectrum flying operations and deployments globally. This also includes theatre munitions movements, global command and control communications to forward deployed locations, support for theatre intelligence operations, and joint/combined training.

RAF Alconbury was controlled by the USAAF Eighth Air Force during the Second World War, but after war ended in 1945, it became the US Strategic Air Forces in Europe (USSAFE), thereafter, the USAFE. It was in 1938 when No. 63 Squadron, flying the Fairey Battle light bomber, arrived to RAF Alconbury; during this time, the airfield consisted of a few wooden huts.

In September 1939, RAF Alconbury became RAF Wytons satellite airfield under the control of No. 2 Group with three squadrons. These were regularly deployed to Alconbury. In April 1940, No. 15 Squadron arrived at RAF Alconbury, and it flew its first raid of the war in May of that year against a German airfield. The next operation was dreadful as only six out of the twelve aircraft that flew the mission returned to base.

The remaining aircraft of No. 15 then moved back to RAF Wyton, and Alconbury reverted to satellite use by both Wyton squadrons. In the autumn of 1940, units were converted to Vickers Wellington bombers, and in November 1940, RAF Wyton and Alconbury came under the control of No. 3 Group. In the early months of 1941, the expansion of RAF Alconbury commenced to upgrade its facilities from a satellite airfield to a fully operational one. A main concrete runway was built, and the perimeter track served hard standings. In February 1941, No. 40 Squadron brought its Wellingtons to Alconbury and began to operate night raids against industrial targets in Germany. These raids continued until the autumn of that year.

In the winter of 1941, sixteen Wellington bombers were deployed to Malta, but by February 1942, it was evident that the major section of No. 40 would not be returning from the Mediterranean area and the remaining aircraft at RAF Alconbury formed into No. 156 Squadron RAF.

Operations from Alconbury with No. 3 Group continued until August 1942 when No. 156 was chosen to become one of the special Pathfinder Force units and moved to RAF Warboys later that same month. This was the end of RAF Bomber Command's association with Alconbury. During the RAF Bomber Command operations flown from RAF Alconbury, over sixty-seven bombers were destroyed, of which eight were Blenheims and fifty-nine were Wellingtons.

After the terrible attacks on the US Navy at Pearl Harbor, and the declaration of war by Germany against the US in December 1941, the governments of the UK and US decided that the main Allied effort was to be against Nazi Germany first, while containing the Japanese in the Pacific.

In 1942, RAF Alconbury was transferred from RAF Bomber Command to USSAF control, with several RAF stations in the East Anglia area handed over to the US. The first USAAF unit to be activated at Alconbury was the 357th Air Services Squadron in August 1942. The first base commander was Col. Edward J. Timberlake, taking command in December 1942.

The 93rd Bombardment Group (BG) was the first unit to occupy RAF Alconbury in 1942. Also known as the 'Travelling Circus' from Fort Myers, Florida, the 93rd BG was assigned to the 20th Combat Bombardment Wing (CBW) at RAF Horsham, Norwich, flying the B-24 Liberator.

The 93rd B-24 Liberator Bomber Group was assigned to attack steel and engineering works at Lille, France. Until December, the group operated primarily against submarine pens along the French coast along the Bay of Biscay. His Majesty King George VI paid his first visit to an Eighth Air Force base in November 1942, and during the visit, he was shown the B-24 *Teggie Ann*, then considered to be the 93rd's leading aircraft.

In December 1942, the group was transferred to Twelfth Air Force in North Africa to support the Operation Torch landings. January 1943 saw the arrival of the 92nd Bombardment Group to RAF Alconbury, equipped with the Boeing B-17 Flying Fortress. The 92nd BG was known as the 'Fames Fews', and it was assigned to the 4th Combat Wing at RAF Thurleigh. The 92nd was withdrawn from combat and its B-17F bombers were exchanged for the older version of the B-17E bombers, which were flown by the 97th BG. The 92nd BG then acted as operational training unit, supplying combat crews to groups in the UK. In the early months of 1943, the 92nd BG was to return to combat operations. This was due to the British-American invasion of French North Africa during the North African Campaign known as Operation Torch.

In May 1943, the 92nd BG began flying missions engaging strategic targets including shipyards, ball-bearing plants, submarine installations, airfields, aircraft factory, and a magnesium mine and reducing plant in Norway. Later that same year, the 92nd was moved to RAF Podington when RAF Alconbury was removed from operational bombing missions to the Pathfinder missions with the 482nd and 801st Bomb Groups. Additionally, 1943 saw twenty-two B-17F Flying Fortress bombers modify to the YB-40 configuration and deployed to RAF Alconbury for further testing and evaluation. The first operations took place in early summer of 1943 against U-boats at Saint-Nazaire in France. During operations, the YB-40 project was recognised as an operational failure, with many standards not being met while on sorties. The YB-40s were converted back to the B-17F and returned to the States and used as training aircraft.

August 1943 saw 482nd Bomb Group form at RAF Alconbury with the 812th, 813th, and 814th Bombardment Squadrons. In September 1943, the 482nd Bomb Group was activated, which was unique as this was the only bomb group to be activated in the UK from scratch. In late 1943, two new bomb squadrons were formed: the 36th and 406th, which were equipped with specially modified Consolidated B-24 Liberators and attached to the 482nd BG. Operation Carpetbagger began soon after.

In March 1944, the 482nd BG was removed from combat operations and became a training unit, but also continued special operations, especially during D-Day. The 482nd BG continued training operations, graduating over thirty-six navigators each month. From August 1944 to April 1945, the 482nd BG administered over 200 radar operations over hostile territory, dropping over 40 tons of bombs in Germany.

The object of the Carpetbagger project was to fly Special Operations missions that delivered supplies to resistance groups in enemy occupied countries. The squadrons flew agents and supplies into southern France with B-24 Liberators that had all armament removed except in the top and tail turrets. The undersides of the aircraft were painted black to avoid detection by enemy searchlights.

In 1944, many bomb groups were relocated to other bases around the UK and this continued during 1945, but RAF Alconbury carried its operational status with many missions being flown. In addition to be an operational bomber base, RAF Alconbury also became a major maintenance base, which was under the command of the 2nd Strategic Air Depot based at RAF Abbots Ripton.

The Grange Hotel in Brampton, located south-west of Alconbury, became the command and control faculty by the USAAF, and in June 1942, it became the HQ of the 1st Bombardment Wing. The 1st Bombardment Wing were renamed the 1st Bombardment Division in 1943 and renamed again to 1st Air Division when they became an operational combat wing in 1945. Operational missions finished at RAF Alconbury in April 1945, and the 4822nd Bomb Group departed back to the States in May that year. The 36th Bomb Group was deactivated in the winter of 1945.

The 857th Bombardment Squadron from the 492nd Bombardment Group was transferred to Alconbury in June 1945 from RAF Harrington after the closure of the airfield. The 857th used its B-24s for various cargo-ferrying operations to and from the continent until being deactivated. The HQ of the 1st Air Division was transferred to RAF Alconbury in September 1945 due to the closure of the Grange Hotel. With the 1st Division deactivated in October 1945, RAF Alconbury was transferred back to RAF command in November 1945. Both Alconbury and Abbots Ripon were placed under care and maintenance status, used by the RAF for storage, distribution, and the dismantling of munitions.

With growing tensions and threats by the Soviet Union in 1948 and with the Berlin Blockade and the invasion of South Korea, it was decided in 1951 to re-establish a strong American presence in Europe. RAF Alconbury was designated as Operating Location 1 for the Third Air Force in June 1952. The semi-permanent force became fully permanent a year later when the USAFE was officially given command over both the Alconbury Airfield and Abbots Ripton depot facilities.

RAF Alconbury suffered insufficient facilities from the Second World War and needed major upgrading and expanding. It was planned to design a major expansion to accommodate the new jet aircraft and other operational facilities. A new control tower and seventy-five new aircraft hardstands were planned. The upgrading, strengthening, and extension of the runway was also required. Numerous support buildings, barracks, and other facilities were constructed during 1953 to 1955. The most dramatic development was the construction of a new bomb dump to replace the small existing site.

By 1954, a large area known as the 'Peninsular site' was constructed. It consisted of up to fifteen hexagonal watchtowers, plus Igloo-style buildings, both of which were designed to hold nuclear weapons, although none were believed to be stored at the base during its operational lifetime. Several other structures were built, including a new boiler house for the aircraft maintenance support area, a new US Navy-style hangar, and the new airfield control tower, which replaced the smaller, two-storey wartime towers. In March 1955, construction was completed to expand the command organisation at Alconbury to a group level, with the 7560th Air Base Group (7560th ABG) being activated. The group's mission was to support tactical and logistical units assigned to the base. Other units were the 7560th Air Base Squadron, 7560th Air Police Squadron, and 7560th Materiel Squadron.

Command of RAF Alconbury was transferred to RAF Sculthorpe in September 1955. The 7560th was activated in January 1956, and in May 1957, Alconbury was designated a primary base. It was not until 1955 when RAF Alconbury became a USAFE flying unit again, activating after ten years of non-flying duties.

The 86th Bombardment Squadron, flying the B-45A Tornado, began operations in 1955 at RAF Alconbury as a detachment of the Tactical Air Commands 47th Bombardment Wing, which was stationed at RAF Sculthorpe. The 47th Bombardment Wing began flying into RAF Alconbury in 1958 with its newly equipped B-66 Destroyers, which were to replace the B-45s. With the new equipment, the 86th was redesignated the 86th Bombardment Squadron, and both the 47th Bomb Wing and 86th Bomb Squadron made up part of the Tactical Air Command.

In April 1959, the 53rd Weather Reconnaissance Squadron arrived from RAF Burtonwood, following the closure of Burtonwood's runway and drawdown of the USAF presence there. Flying the WB-50D Superfortress, the 53rd's missions were to collect weather data that would be transmitted to weather stations preparing forecast for the air force. After a brief occupancy at RAF Alconbury, the squadron was reassigned to its permanent base at Mildenhall in August 1959. In that same year, the 10th Tactical Reconnaissance Wing (10th TRW) arrived from Spangdahlem Air Base, Germany, to replace the 7560th Air Base Group as the host unit at RAF Alconbury. The 86th Bomb Squadron was returned to RAF Sculthorpe. Command was transferred from the 49th Air Division to the Third Air Force, with RAF Alconbury hosting the 10th ARW for next three decades.

Further construction was required with the arrival of the 10th TRW and the B-66s. Nose docks were constructed to provide protection from the weather for the delicate electronics in the B-66s that were parked outside on hardstands, as well as for the cockpit, engines, and radar. The 10th TRW's HQ was located at RAF Alconbury; however, due to space limitations, the wing's component squadrons were dispersed. The wing consisted of the 1st and 30th Tactical Reconnaissance Squadrons (TRS), which were based at Alconbury. These squadrons flew the RB-66B Destroyer, a day/night photographic reconnaissance aircraft.

The closure of Bruntingthorpe in 1962 and the closure of the active runway at Chelveston in 1963 saw the 19th and 42nd TRSs transfer to Toul-Rosieres Air Base, France, where they operated for a few years as Detachment 1, 10th Tactical Reconnaissance Wing. Eventually, the 10th TRW rotated four of its squadrons to Toul-Rosieres Air Base, having two at Alconbury and two others forward deployed in France.

At RAF Alconbury, the 10th TRW retained the 1st and 30th TRSs with their RB-66Bs until May 1965, when it began conversion to the RF-4C Phantom II, the unarmed photographic reconnaissance version of the USAF's F-4C fighter. The wing added a third RF-4C squadron in August 1966, with the addition of the 32d TRS. This squadron had formerly flown RF-101 Voodoos with the 66th TRW at Laon-Couvron Air Base, France.

Following the closure of USAF bases in France in 1967 and rising tension in Libya, a major repositioning of USAFE assets in the UK was conducted. In 1970, RAF Wethersfield became a standby deployment base and was placed under the command of Alconbury. In 1972, RAF Alconbury runway was overhauled due to the number of F-4 Phantoms using the runway, and the aircraft of the three squadrons of the 10th TRW went to RAF Wethersfield to fly out their sorties. Missions from this base were highly successful, due to the diligence and demanding work of all temporarily assigned personnel. Wethersfield remained a satellite base for RAF Alconbury until July 1990, when it was closed and handed back to the RAF.

By the mid-1970s, the need for tactical reconnaissance aircraft was drastically reduced, and budget reductions saw the numbers of front-line tactical reconnaissance aircraft decreased. In 1976, two of the 10th TRW's squadrons (32nd TRS and 30th TRS) were deactivated. The 1st TRS remained the only squadron providing battlefield tactical reconnaissance.

In April 1976, the 10th TRW was chosen to become the USAFE's aggressor training unit. The 527th Tactical Fighter Training and Aggressor Squadron was activated and equipped with the F-5E in May 1976. The aircraft were originally part of an order for South Vietnam. The 527th began providing aggressor support to European-based combat units in September. It was subsequently renamed as the 527th Aggressor Squadron in 1983.

The 527th conducted its training mission both from RAF Alconbury and would deploy frequently to USAF and NATO bases within Europe. By 1979, a large training area over the Mediterranean was established with a large facility established at Decimomannu Air Base on the island of Sardinia, Italy.

In 1988, the fleet of aggressor F-5Es of the 527th Aggressor Squadron was getting rather worn out because of sustained exposure to the rigours of air combat manoeuvring. It was decided to re-equip the squadron with F-16C Fighting Falcons and reassign the squadron to RAF Bentwaters. With this, the A-10s at Bentwaters would be reassigned to Alconbury, providing the 10th with a new Close Air Support (CAS) mission.

The 527th AS flew its last F-5E sortie from Alconbury in June 1988. In July 1988, the squadron began re-equipping with F-16Cs, which was completed by mid-January 1989 at Bentwaters. Yet in 1990, with the collapse of the Soviet Union, it was decided to terminate the entire USAF aggressor program. The 527th AS was deactivated in late autumn of 1990.

In October 1990, the 17th Reconnaissance Wing (RW) was activated. The 17th RW was assigned to SAC Eighth Air Force, 7th Air Division. The 95th Reconnaissance Squadron was the operational squadron flying the TR-1A a tactical version on the Lockheed U-2.

With the arrival of the U-2, RAF Alconbury began specialised construction. Work included the construction of five prefabricated 'Ready Sheds', thirteen extra-wide hardened aircraft shelters, a squadron headquarters, a massive Avionics and Photography Interpretation Centre, and new concrete aprons and taxiways. In addition, to ensure that the 17th RW would have a command post for its aircraft, a command post facility was constructed with its own power plant, communications facilities, air supply, and decontamination facility, to help the wing continue operations during a nuclear war.

RF-4 Phantoms were also becoming much more expensive to maintain, and in July 1987, the RF-4Cs of the 1st TRS flew its last mission and the squadron was deactivated in January 1988. Some of its aircraft were sent to the 26th TRW at Zweibrücken Air Base, West Germany, while the rest went to Air National Guard units as replacement aircraft or to AMARC for storage.

McDonnell Douglas F-4 Phantom at Volkel Luchtmachtdagen 2013 on its farewell tour.

The Lockheed U-2, nicknamed *Dragon Lady*, parked up gracefully at RAF Fairford.

After the withdrawal of the RF-4Cs and F-5Es, the 10th TRW became the 10th Tactical Fighter Wing (TFW) in August 1987, with two squadrons of A-10A aircraft. The 509th and 511th Tactical Fighter squadrons (TFS) were assigned to the 10th TFW, and in June and September 1988, both squadrons relocated to the 81st Tactical Fighter Wing at RAF Bentwaters.

Four squadrons equipped with the A-10 'Warhog' arrived and were assigned to RAF Bentwaters. It was decided that with the deactivation of the RF-4Cs at RAF Alconbury, two of the squadrons would be relocated there in a dispersal move, with the other two remaining at Bentwaters.

Some of the first aircraft to be sent to the Gulf Area during Operation Desert Storm were three TR-1As from RAF Alconbury, deployed to Taif Air Base in Saudi Arabia in August 1990. In addition, over twenty A-10As of the 511th TFS were deployed to Damman/King Fahd International Airport Saudi Arabia in December 1990 as part of the 354th TFW, which was deployed from Myrtle Beach Air Force Base, South Carolina. The squadron's aircraft arrived the next day.

The 511th TFS flew more than 1,000 combat missions during Operation Desert Storm and destroyed nearly 200 tanks, 100 armoured personnel carriers, 200 artillery sites, seventy-six radar sites, thirteen surface-to-air missile sites, twelve Scuds, and other military equipment.

With the end of Gulf War, all the aircraft had returned to RAF Alconbury by June 1991 and the 17th RW was deactivated, but its secondary unit, the 95th RS, remained at RAF Alconbury. It subsequently deactivated at Alconbury in September 1993, then reactivated in 1994 as the 95th RS at RAF Mildenhall, assigned to the 55th Operations Group. The squadron provides intelligence support to produce politically sensitive real-time intelligence data vital to national foreign policy.

In December 1991, the 509th TFS flew its last operational mission. The 511th TFS's last mission was in March 1992. Throughout 1992, the 10th TFW's A-10 aircraft were transferred back to the US. The 509th TFS's aircraft were sent directly to AMARC for long-term flyable storage. Some of the 511th TFS's aircraft were sent to Air National Guard units, the remainder to AMARC storage. The last aircraft departed the Alconbury runway in December 1992. Both fighter squadrons were deactivated in that month.

In December 1992, the 39th Special Operations Wing arrived at RAF Alconbury, consolidating its units from RAF Woodbridge and Rhein Main Air Base, Germany. After consolidating its aircraft and people at the base, the 39th SOW deactivated, and the 352nd Special Operations Group activated, linking the unit's heritage with a historic Second World War commando unit.

In May 1993, the drawdown of American forces in Europe began, and it was announced that operational duties at RAF Alconbury would be reduced. The 10th Air Base Wing was deactivated in October 1994. In its place, the 710th Air Base Wing (ABW) was activated as the host unit on RAF Alconbury.

The 352nd Special Operations Group and its associated aircraft transferred to RAF Mildenhall in February 1995, which ended active USAF flying operations at RAF Alconbury. The airfield returned to the MOD in September 1995. The main base support areas, such as housing, Base Exchange, commissary, financial institutions, administrative, and support offices, were retained under USAF control.

Fairchild A-10C Thunderbolt II 82-0648/82-648/DM from the 354th Fighter Squadron at Davis-Monthan Air Force Base, Arizona.

Fairchild A-10C Thunderbolt II 81-976 from 52nd Fighter Wing, 81st Fighter Squadron, Spangdahlem Air Base, Germany.

In July 1995, the 710th Air Base Wing was deactivated and the 423rd Air Base Squadron at RAF Molesworth assumed the host unit role at RAF Alconbury as well as RAF Upwood. In July 2005, the squadron was redesignated as the 423rd Air Base Group and its headquarters and mission was moved to RAF Alconbury.

The 501st Combat Support Wing (501 CSW) was reactivated in March 2005 at RAF Mildenhall. Its mission was administering the various geographically separated units in the UK, and in May 2007, the wing moved to RAF Alconbury, where the wing still occupies at present.

2
Operational Air Bases: RAF Mildenhall, RAF Lakenheath, RAF Fairford

RAF Mildenhall

RAF Mildenhall is a RAF station located near Mildenhall in Suffolk, England.

The base primarily supports USAF operations, and is currently the home of the 100th Air Refuelling Wing (ARW). RAF Mildenhall houses units from four different major USAF commands—Air Combat Command, Air Force Special Operations Command, Air Mobility Command, and USAFE—and units from the US Navy. The main host unit based at RAF Mildenhall is the 100th ARW, managing aircraft from the European Tanker Task Force, which provides air refuelling to US and NATO allies' aircraft in European operations. It is the only permanent USAF air refuelling wing in Europe and was activated at RAF Mildenhall in February 1992. The 100th ARW provides the critical air refuelling 'bridge' that allows the Expeditionary Air Force to deploy globally with the appropriate urgency.

The 352nd Special Operations Wing (SOW) is the air force component for Special Operations Command Europe. It transferred to RAF Mildenhall in February 1995, from RAF Alconbury. The 352nd SOW mission statement is to serve as a focal point for all USAF special operations activities throughout Europe, as well Africa and the Middle East. The 352nd SOW conducts a variety of high-priority, low-visibility missions supporting US and allied special operations forces.

The 95th Reconnaissance Squadron (RS) conducts RC-135 Rivet joint flight operations in Europe and the Mediterranean as tasked by National Command Authorities and European Command. It was activated at RAF Mildenhall in July 1994, having been previously assigned to RAF Alconbury.

RAF Mildenhall opened in 1934, and King George V reviewed 350 aircraft there in 1935 on his Silver Jubilee. This historical event is commemorated by a memorial located in front of the Building 562, the current 100th ARW HQ. During the Second World War, RAF Bomber Command used the station for operational combat missions. A variety of aircraft was flown during this period, including Vickers Wellingtons, Short Stirlings, and Avro Lancasters.

After the war, RAF Mildenhall was placed on standby but was reopened by the RAF in July 1950, becoming a USAF–RAF joint operation base. Assigned to SAC to station B-29 Superfortress bombers, it became a B-50 Superfortress base in 1952, and a B-47 Stratojet and KC-97 Stratofreighter base from 1953 until 1958.

Above: F-16C 89-0023 from the Turkish Air Force arrives at RAF Mildenhall during a transit flight back from the States heading to Turkey.

Below: A-10C Warthog 81-0964 from Spangdahlem Air Base, heading home to Germany after being on deployment in the States.

RAF Mildenhall closed for runway repairs throughout 1958 but reopened again in March 1959 when the Military Air Transport Service transferred its main UK terminal to Mildenhall from RAF Burtonwood. The base has become 'The Gateway to the United Kingdom' for most US military personnel and dependants arrive or depart the UK from RAF Mildenhall. In September 1959, RAF Mildenhall was assigned to USAFE, and the RAF relinquished joint operations status. This has been in continuous USAFE operation to the present day.

In 1978, the 306th Strategic Wing was transferred to RAF Mildenhall from Ramstein Air Base, West Germany, with a KC-135 air refuelling and a RC-135 reconnaissance mission. The 306th also functioned as the focal point for all operations in Europe and became the host unit at RAF Mildenhall. RAF Mildenhall became the European Reconnaissance centre, and for many years, several types of Boeing RC-135 reconnaissance aircraft regularly arrived and departed from Mildenhall. Most of these aircraft came from the 55th Strategic Reconnaissance Wing at Offutt AFB, Nebraska, and had the capability to receive radar and radio signals from far behind the borders of the Communist Eastern Bloc. From Mildenhall, the RC-135s flew ELINT and COMINT missions along the borders of Poland, the Soviet Union, and Czechoslovakia. The arrival of Detachment 4, 9th Strategic Reconnaissance Wing, in 1976 saw rotational Lockheed U-2 and SR-71 Blackbird operations from the base.

These aircraft carried out strategic photo reconnaissance missions for NATO allies and the USAF within the framework of the SALT I Agreement of 1972. Under this agreement, the Soviet Union and the US reached agreement on a partial freeze on the number of offensive nuclear weapons, and these flights were to check that the Soviets were adhering to the agreement.

The 306th also played a significant role in the success of Operation El Dorado Canyon—the American attack on the Libyan capital of Tripoli in 1986. In support of this fourteen-hour, radio-silence rendezvous mission, the unit deployed the largest number of refuelling aircraft ever flown over Europe, and the largest fleet of KC-10s ever airborne at one time. From its arrival until the departure of the last SR-71 in January 1990, the 306th Strategic Wing's SR-71 and U-2 aircraft came to symbolise RAF Mildenhall in the local resident's eyes. The 306th operated at RAF Mildenhall until February 1992, when the 100th ARW of USAFE assumed the 306th SW's responsibilities, and became host unit at Mildenhall, along with becoming HQ European Tanker Task Force with its KC-135 Stratotankers.

The 100th ARW provided aerial refuelling support to Air Force, Navy, and Marine Corps' aircraft, as well as aircraft of allied nations. Its aircraft are also capable of transporting litter and ambulatory patients using patient support pallets during aeromedical evacuations.

In May 1993, as part of the drawdown of forces in Europe, the 352nd Special Operations Group and its associated aircraft—the MC-130E, HC-130P/N, and MH-53—transferred to RAF Mildenhall in March 1995.

In July 1994, the 95th RS was activated at Mildenhall, essentially taking over the mission of the deactivated 306th Strategic Wing.

In January 2015, the US Department of Defense announced the closure of RAF Mildenhall, and absorbed 700 personnel in Ramstein Air Base and Spangdahlem Air Base. The changes were a result of the US Department of Defense's European Infrastructure Consolidation (EIC)—a two-year-long study to downsize the number of sites and facilities being used by US forces in Europe. On the same day, the UK government announced the site was one of twelve

Lockheed SR-71 *Blackbird* is a long-range, Mach 3+ strategic reconnaissance aircraft, displayed here at RAF Fairford in 1991.

Above: Lockheed U-2 *Dragon Lady* lifts off from RAF Fairford in the early hours of the morning, heading back to the US.

Below: Two Boeing KC-135 Stratotankers parked on the ramp in the summer at RAF Mildenhall in the 1988.

Lockheed MC-130H Combat Talon II 87-0024 departing RAF Mildenhall on a training mission in the late afternoon.

that would be sold as part of the strategy for the MOD estate, although the date for the sale was given, this has recently been changed to 2024. However, a change of administration in the US heightened security in Europe and concerns about the Middle East have prompted a rethink on Mildenhall's closure.

RAF Lakenheath

RAF Lakenheath is located near the town of Lakenheath in Suffolk. The base supports the USAF operations and is home of the 48th Fighter Wing (FW).

The 48th is one of the primary fighter wings of the USAFE and has been part of it for over fifty years. The 48th FW has over 5,500 active military personnel and over 2,000 British and US civilian contractors. The 48th and RAF Lakenheath is home to three squadrons of McDonnell Douglas F-15 Eagle fighter jets. These jets are the only USAF F-15 fighters in the UK and Europe. In addition to supporting three combat-ready squadrons, the 48th formerly housed the 56th Rescue Squadron (56 RQS), which operates HH-60G Pave Hawk helicopters, conducting search and rescue missions. The squadron relocated to Aviano Air Base, Italy in June 2018.

Lakenheath Warren was first used during the First World War with the area being used by the RFC for bombing and ground-attack training. The airfield had little activity and was abandoned once the war had ended in 1918. Four months into the Second World War, the Air Ministry decided to select Lakenheath as the alternative airfield for RAF Mildenhall as a decoy, using false lights, runways, and aircraft to divert Luftwaffe attacks away from Mildenhall.

In 1941, it was decided that runways and hardstands for thirty-five aircraft, along with two hangers, were to be built, and the RAF started using the airfield later that year. Lakenheath started functioning as a satellite airfield for Mildenhall with Short Stirling bombers from

Above: F-15E Strike Eagle 01-2000 from the 494 Fighter Squadron departs RAF Lakenheath on a training exercise with full afterburner.

Below: Sikorsky HH-60G Pave Hawk 91-26352 arriving back to base after a general handling sortie.

149 Squadron. In late 1941, the squadron changed to the Vickers Wellingtons, and once they became fully operational with their new aircraft, the squadron moved to their new permanent home at Lakenheath and remained there until mid-1944, when the squadron moved to RAF Methwold.

In early 1943, three more hangars were erected on the north side of the airfield for glider storage, but the gliders were dispersed at Lakenheath during that year. Also, in 1943, a second Stirling squadron moved to Lakenheath but transferred to No. 100 Group RAF for bomber support, moving to RAF North Creake in 1944. No. 149 Squadron also ended its association with RAF Lakenheath the same year, taking its Stirlings to RAF Methwold.

In 1945, Lakenheath was selected to be upgraded to a Very Heavy Bomber airfield. Lakenheath was one of three RAF airfields being prepared to receive USAAF Boeing B-29 Superfortresses, planning to replace some of Eighth Air Force's Third Air Division's Consolidated B-24 Liberator groups.

Work began to remove the existing runways and lay new ones comprising of high-grade concrete. The main runway, 07/25, was 2,743.2 metres long, and runways 01/19 and 14/32 were both 1,828.8 metres long; all three were 91.44 metres wide. During the construction, over 1,000 men were working on the site and it was expected to be completed within twelve months, but this overran extensively and took over two and half years to finish, costing nearly £2 million. By the time construction had been completed, the Second Word War had ended, and RAF Lakenheath was put on a care and maintenance status.

In 1947, the RAF returned to Lakenheath and the runways were repaired, resurfaced, and readied for operations by mid-1948. July 1948 saw Boeing B-29 Superfortresses of the SAC's 2nd Bombardment Group deployed to Lakenheath for a ninety-day deployment. In November 1948, the transfer of operational control from the RAF to the USAFE was granted; the unit at RAF Lakenheath was the 7504th Base Completion Squadron. The squadron was elevated to an air base group (ABG) in January 1950 and to a wing (ABW) in September 1950.

By 1950, Lakenheath was one of three main operating bases for the US SAC in the UK, and in 1951, Lakenheath was transferred from USAFE to SAC, and placed under the 3909th ABG. In 1952, it was decided that high-security perimeter fencing was to be erected around the base.

A series of SAC, bomber squadrons, and wings rotated through RAF Lakenheath between 1949 and 1956 with the B-29s giving way to the improved B-50 Superfortresses and B-47 Stratojet. It was not until 1960 that the first fighter jet arrived at RAF Lakenheath. This was partly because French President Charles de Gaulle insisted in 1959 that all non-French nuclear-capable forces should be withdrawn from his country.

During the period from 1960 to the present date, RAF Lakenheath has experienced many aircraft types including the McDonnell Douglas F-4D Phantom II, General Dynamics F-111As, and, finally, the McDonnell Douglas F-15 Strike Eagles, which arrived in 1992 and are still present today alongside the HH-60G Pave Hawk helicopters.

The General Dynamics F-111 Aardvark departs RAF Alconbury with full reheat.

F-15C Eagle 86-0172 from the 493rd Fighter Squadron takes off from RAF Lakenheath with full reheat heading out on another mission

RAF Fairford

RAF Fairford is a RAF station, located in Gloucestershire, which is currently used as a standby airbase and not in everyday use.

It has been used as an airbase for USAF B-52s during the 2003 Iraq War, Operation Allied Force in 1999, and the First Gulf War in 1991. It is the USAF's only European airfield for heavy bombers and is still used today by the USAF. The airfield has also been occasionally used by American B-2 Spirit stealth bombers and is frequently visited by transient U-2 aircraft.

RAF Fairford was the only Transoceanic Abort Landing site for NASA's Space Shuttle in the UK. As well as having a sufficiently long runway for a shuttle landing, it also had NASA-trained fire and medical crews stationed on the airfield. The runway is rated with an unrestricted load-bearing capacity, meaning that it can support any aircraft with any type of load.

RAF Fairford was built in 1944 and this was to serve the British and American troops for the D-Day invasion of Normandy during the Second World War. In the 1950s, at the beginning of the Cold War, RAF Fairford was transferred to the USAF for strategic bomber operations; for this purpose, a new 3,000-metre runway was constructed and completed in 1953, and, to this day, it is still one of the largest runways in Britain.

In 1969, Fairford was chosen as one of the airfields for Concorde to undergo its flight testing, which continued until 1977. In November 1978, the 11th Strategic Group was activated at RAF Fairford but was unmanned until the following February. It used the KC-135 aircraft and crews from the Air National Guard and AF Reserve units until the 11th SG received its own aircraft in 1979. It soon began aerial refuelling support for all USAF operations, deployments, and redeployments, as well as participating in NATO exercises.

Due to RAF Fairford's location and infrastructure, the airfield is a designated forward operating location for the USAF. RAF Fairford was used during the First Gulf War back in 1991, with the Gloucestershire skies having the sights of the Boeing B-52 Stratofortress and the Boeing KC-135 Stratotanker. Fairford was again used in 1999 when B-52s from Barksdale Air Force Base, Rockwell B-1 Lancers from Ellsworth Air Force Base, and KC-135s

Lockheed U-2 *Dragon Lady* arriving at RAF Fairford on its homeward journey back to Beale Air Force Base.

Above: Lockheed U-2 *Dragon Lady* taxies off the active runway at RAF Fairford after completing the first leg of the journey home to the US. Piloted by Maj. Zachary Johnson, 99th Reconnaissance Squadron.

Below: Lockheed U-2 *Dragon Lady* taxies onto the active runway at Fairford for the homeward journey to Beale Air Force Base.

Above: Two Boeing B-52 Stratofortresses begin their taxiing to the active runway at RAF Fairford for a mission during Exercise BALTOPS, 2015

Below: Boeing B-52 Stratofortress departs RAF Fairford on a morning mission during Exercise BALTOPS, 2015.

from Mountain Home AFB were operational. In 2003, during the Iraq War, Operation Iraqi Freedom included more B-52s based out of Minot Air Force Base.

During recent years, RAF Fairford has undergone a lot of redevelopment work on both runway and infrastructure; this included the construction of two climate-controlled hangars for B-2 stealth bombers and a low-observability maintenance dock.

It was announced that the USAF would withdraw all their uniformed staff from the station by September 2010, after which it would be run on a 'care and maintenance' basis. However, the base would remain a designated satellite airfield, capable of immediate reactivation within twenty-four to forty-eight hours. As part of the airframe rotation program, RAF Fairford remains the fuel and rest period airfield for the US reconnaissance aircraft U.2 *Dragon Lady*, based at Beale Air Force Base in America. The rotation program normally takes place every two to three months, with aircraft heading to the Middle East to support US operations or returning to the States.

The normally dormant RAF Fairford has become active in the past couple of years in preparation for hosting long-range bombers from the USAF on deployments. RAF Fairford saw President Barack Obama arrive on 3 September 2014 during a NATO Summit in Newport in the UK.

RAF Fairford became active in June 2015 when three USAF Boeing B-52H Stratofortress bombers were assigned to the 5th Bomb Wing at Minot Air Force. Additionally, two B-2A Spirits were deployed from Whiteman AFB, Missouri, during Operation BALTOPS 2015, the largest multinational exercise taking place in the Baltic Sea. It was hosted by the Commander US Naval Forces Europe, testing allied and partner maritime niche capabilities including maritime security tasks.

June 2016 marked the beginning of a now annual deployment from the USAF Global Strike Command (BALTOPS 2016), which again was hosted at RAF Fairford. Returning this year, three B-52H Stratofortresses arrived from the 5th Bomb Wing based at Minot Air Force Base, North Dakota.

For the second time in 2016, RAF Fairford became active for the preparation of hosting long-range bombers from the USAF with the arrival of two Rockwell B-1 Lancer bombers from Dyess Air Force Base, Texas, and one Boeing B-52H Stratofortress operating under the 307th Bomb Wing from Barksdale Air Force Base, Louisiana.

Ample Strike is a multi-national Close Air Support (CAS) exercise specifically designed to train and exercise Joint Terminal Attack Controllers (JTAC). Hosted by the Czech Republic, additional participation was built up of a further fifteen NATO and partner nations from Belgium, Croatia, Estonia, Finland, France, Germany, Great Britain, Greece, Hungary, Lithuania, Latvia, the Netherlands, Poland, Slovenia, and the United States. Many of these nations deployed units and operated directly from military bases throughout the Czech Republic.

June 2017 saw the arrival of two B-2 Spiritbombers from Whiteman AFB, Missouri. These were joined by three B-1B Lancers from Ellsworth AFB, South Dakota, and three B-52H Stratofortresses from Barksdale Air Force Base, Louisiana, to participate in Saber Strike and Baltic Operations in the US-European Theatre.

RAF Fairford is home of the Royal International Air Tattoo (RIAT), boasted to be the world's largest military air show, held annually in July in support of the Royal Air Force Charitable Trust. The show typically attracts a total of 150,000 to 160,000 spectators over the weekend. The 501st Combat Support Wing offers support during this event alongside civilians.

Above: Rockwell B-1B Lancer 85-084 departs RAF Fairford on an early morning mission during Exercise Sabre, 2017.

Below: Rockwell B-1B Lancer 85-084 taxies to the active runway at RAF Fairford on an early morning mission during Exercise Sabre, 2017.

Above left: Northrop Grumman B-2A-30 Spirit 88-0329 climbs out of RAF Fairford. heading to Whiteman Air Force Base.

Above right: Northrop Grumman B-2 Spirit 82-1068 *Spirit of New York* taxies to the ramp at RAF Fairford after arriving from Whiteman Air Force Base.

The Lockheed U-2 *Dragon Lady* arriving at RAF Fairford on a rotational mission heading to the Middle East.

Ground crew connect the wheel stabilisers on the U-2 before it can taxi to parking after an early morning arrival at RAF Fairford.

Boeing VC-25 92-29000 *Air Force One* arrives at RAF Fairford for the NATO Summit, 2014.

Above: Air Force One taxies to the parking dispersal at RAF Fairford during the NATO Summit in 2014.

Below: President Barack Obama walks down the aircraft stairs to his awaiting entourage during the NATO Summit, 2014.

Boeing B-52H Stratofortress 60-0037 sits peacefully on the tarmac at RAF Fairford during Exercise BALTOPS, 2015.

Aircrew from the 5th Bomb Wing, based at Minot Air Force Base, head back for their briefing after another successful mission at RAF Fairford.

Above: Boeing B-52H Stratofortresses from 5th Bomb Wing, Minot Air Force Base, departs RAF Fairford during Exercise BALTOPS, 2016.

Below: Boeing B-52H Stratofortress 60-0047 from 5th Bomb Wing, Minot Air Force Base, parked up on the ramp at RAF Fairford.

Two Boeing B-52 Stratofortresses depart RAF Fairford for a mission during Exercise BALTOPS, 2016, trailing smoke as they leave.

The Boeing B-52 Stratofortress aircrew are ready to taxi for another mission during Exercise BALTOPS at RAF Fairford.

Above: Rockwell B-1B Lancer 85-089 departs RAF Fairford on a morning mission during Exercise Ample Strike, 2016.

Below: Colonel Denis Heinz, commander of the 489th Bomb Group, Dyess Air Force Base, Texas, stands proud with the Rockwell B-1 Lancer aircraft.

Rockwell B-1B Lancer with full reheat takes off from RAF Fairford on another training sortie.

Above: Call sign 'Crook 02', Rockwell B-1B Lancer, taxies of the active runway at RAF Fairford to prepare to take part in Ample Strike, 2016.

Below: The Boeing B-52 Stratofortress taxis across and towards the runway at RAF Fairford for the homeward journey to Barksdale Air Force Base.

Above left: The Boeing B-52 Stratofortress rolling down the runway at RAF Fairford on a mid-morning sortie.

Above right: F-35A Lightning 12-5042 from Luke Air Force Base, with full reheat and vapour during its display routine at RIAT.

Above left: The F-22 Raptor pulling hard to show its topside profile during the RIAT Air Show.

Above right: The Royal Air Force Red Arrows wowing the crowds at RIAT.

3
Operational Wing: RAF Mildenhall

The 100th ARW

The 100th Air Refuelling Wing (ARW) is a USAF unit assigned to the Third Air Force, USAFE-AFAFRICA, and is the only permanent US air refuelling wing in Europe. During the Second World War, the unit's predecessor was the 100th BG from the Eighth Air Force, flying the B-17 Flying Fortress stationed at RAF Thorpe Abbots in England. Flying over 300 combat missions, the group earned two Distinguished Unit Citations. The group suffered tremendous losses in combat, with 177 Aircraft missing in action; it flew its last mission in April 1945. One of the wing's honours is that it is the only modern USAF operational wing allowed to display on its assigned aircraft the tail code (Square-D) of its Second World War predecessor.

The USAFE's only KC-135 air refuelling wing has the responsibility for all the US aerial refuelling operations conducted throughout Europe and supports over 15,000 personnel with over fifteen associated units. June 1943 saw the 100th BG fly is first combat mission against the Bremen U-boat yards—this was the beginning of the 'Bloody Hundredth'.

The group inherited the 'Bloody Hundredth' nickname from other bomb groups due to the amount of losses it took during the war. It was also nicknamed the 'marked outfit' by the German fighters after a B-17 pilot first lowered its landing gear to surrender to three Messerschmitt Bf 109s. The pilot then started to descend after the fighters stopped shooting, then changed his mind and the B-17 gunners shot the three fighter aircraft down.

The 100th BG experienced several instances where it lost a dozen or more aircraft on a single mission, and one raid in October 1943 ended up with only one surviving B-17 Flying Fortress, *The Rosie Riveters*, which returned to Thorpe Abbots. The group received its first DUC after attacking the German aircraft factory at Regensburg in August 1943, resulting in serious disruption to German fighter production. In March 1944, aircrews completed a succession of attacks on Berlin, receiving its second DUC of the war for participating in the Allied campaign against German aircraft factories (Operation Argument) during 'Big Week' in 1944.

During 1944, the 100th BG was involved in conducting missions such as bombing bridges and gun positions to support the invasion of Normandy. Aircrews carried on bombing enemy positions at Saint-Lô and other targets in Brest. The 100th BG attacked enemy and ground defences in the Allied drive on the Siegfried Line, then bombed marshalling yards, German-occupied villages, and communication targets in the Ardennes during the Battle of the Bulge

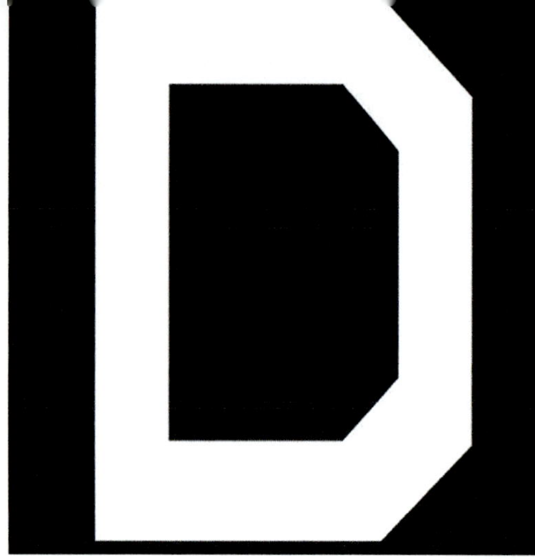

Above: A KC-135R lifts off from RAF Mildenhall on a lunch-time refuelling mission.

Below: Boeing B-17G Flying Fortress 124485 DF-A *Memphis Belle Sally B* flypast at the Duxford Air Museum.

from December 1944 to January 1945. The 100th BG received the French *Croix de Guerre* with Palm for its extraordinary efforts in attacking heavily defended German installations in Germany and dropping supplies to the French Forces of the Interior.

The last combat mission of the Second World War was flown in April 1945, and with this, the crew started dropping food supplies to people in Holland. Later that same year, it transported French Allied former POWs from Austria to France. The group returned to the US and was deactivated in late 1945.

In 1945, RAF Mildenhall saw the first involvement of the USAAF, with two B-17 Flying Fortresses being deployed to conduct bombing trials. A third aircraft shortly followed.

In 1948, the runways at RAF Mildenhall were resurfaced and updated, and in July 1950, it was decided to re-establish a strong American force in Europe. RAF Mildenhall was made available to the USAF by the MOD as a SAC with the B-29 Superfortress to be based there in joint operations with the RAF.

The 100th BW was established in 1953 as part of the SAC, but the wing was activated in 1956 and was assigned to the Eighth Air Force 817 Air Division. The 100th BW was assigned with the new B-47E Stratojet (that had just entered service) and were based from Pease Air Force Base for over ten years. Assigned to preform global strategic bombardment training and air refuelling missions, the 100th BW was deployed to RAF Brize Norton for training purposes.

By the mid-1960s, the B-47 was reaching obsolescence and was being phased out of the SAC strategic arsenal. In October 1965, the USAF initiated Project Fast Fly to oversee the deactivation of the last five B-47 wings and supporting tanker squadrons. The 100th BW was deactivated in June 1966. After its reactivation, the 100th Strategic Reconnaissance Wing performed strategic reconnaissance with the Lockheed U-2 and drone aircraft, and in 1970, it was moved from Biên Hòa to U-Tapao in Thailand.

The Lockheed U-2 *Dragon Lady* taxies to the main runway at RAF Fairford, heading home to Beale Air Force Base.

With the end of US combat operations in Southeast Asia in mid-1973, the USAF formally transferred nuclear air sampling operations to the 100th SRW and the U-2s in South Korea became the 100th SRW. The 100th SRW performed worldwide surveillance missions like the monitoring of the ceasefire between the Israelis and the Egyptians following the 1973 Yom Kippur War. This operation was operated from RAF Akrotiri, Cyprus, and was named Olive Harvest.

In 1976, due to budget cuts, the SAC consolidated its strategic reconnaissance assets. The 99th Strategic Reconnaissance Squadron (SRS) and its U-2s were returned from U-Tapao and assigned to the 9th Strategic Reconnaissance Wing (SRW). With the redesignation, the 100th and its 349th and 350th ARSs were moved administratively to Beale, taking over the assets of the 17th BW, which was deactivated. The 349th and 350th assumed the KC-135s of the 903rd and 922nd ARSs. With the redesignation, the 100th ARW assumed responsibility for providing worldwide air refuelling support for the 9th SRW's SR-71s and U-2s in 1976.

The 100th ARW was activated in 1992. The wing provides critical air refuelling that allows the USAF to deploy around the globe at a moment's notice. The 100th ARW now maintains its own fleet of KC-135s, but it did not receive its first permanent KC-135 until May 1992 and did not reach its full complement of nine tankers until September 1992. The number of permanently assigned KC-135s at RAF Mildenhall reached fifteen aircraft in 1998.

The 100th ARW has supported many US and NATO operations, plus numerous major operations such as Operation Enduring Freedom and Operation Iraqi Freedom. The wing from its base at RAF Mildenhall strived to achieve its role in peacekeeping operations and supporting air operations in Europe, parts of Asia, and Africa.

The year 2017 saw the 100th ARW support many ongoing deployments and missions in the Middle East, Libya, Afghanistan, and Syria. In January, the 100th ARW participated in the operation to refuel two B-2 Spirit from the 509th Bomb Wing that targeted Islamic State training camps in Libya.

In April, the F-35s from Hill Air Force Base participated in a first ever flying training deployment to Europe, and the forward deployment to Estonia, in which the 100th ARW were tasked to refuel on their outgoing routing. Another task in April saw the 100th ARW deploy to Rota Naval Air Station, Spain, to support African Lion, an exercise aimed toward strengthening relationships with US allies and partners.

June saw the 100th ARW supporting exercise BALTOPS at Powidz Air Base, Poland. BALTOPS is an annual multinational, maritime-focused exercise designed to provide training for the participants. Fourteen countries (Belgium, Denmark, Estonia, France, Germany, Latvia, Lithuania, the Netherlands, Norway, Poland, the UK, the US, and NATO's Enhanced Opportunities Partners: Finland and Sweden) participated.

In July, the USAF Thunderbirds were provided with air refuelling. The Thunderbirds were on their way to the 2017 Royal International Air Tattoo air show at RAF Fairford.

The 100th ARW mission goals and vision is 'Ready, Balanced, Better. Square-D Away, This Is "Team Mildenhall" at its finest'.

August 2017 saw the Fleet Logistics Support Squadron 'Rawhides' (VRC-40) based at Norfolk, Virginia, use RAF Mildenhall as their forward-operating base to carry out their missions.

October 2017 saw a formation of KC-135 Stratotankers from the 100th Air Refuelling Wing refuel a B-52 Stratofortress from Barksdale Air Force Base, Louisiana, over the Mediterranean Sea while on deployment during Saber Strike.

Four Boeing KC-135 Stratotankers from the 100th ARW parked on the ramp at RAF Mildenhall.

Above: Two Boeing KC-135 Stratotankers patiently wait for crews to arrive for their next refuelling sortie at RAF Mildenhall.

Below: A KC-135 Stratotanker awaits take-off clearance before heading towards the Middle East.

The Saab JAS 39 Gripen from the Swedish Air Force departing RAF Mildenhall on its way to Red Flag held at Nellis Air Force Base in the US.

Above: The Boeing KC-135 Stratotanker arriving back after a training mission to RAF Mildenhall as dusk begins to draw.

Below: Boeing KC-135R 80-0034 Stratotanker of the 100th ARW departs RAF Mildenhall before a heavy rain shower arrives at the airfield.

Boeing KC-135R Stratotanker 58-0100 taxies to the threshold of runway 29 at RAF Mildenhall.

Boeing KC-135R 61-0304 Stratotanker departs RAF Mildenhall and climbs in the clouds on another refuelling mission.

Above left: Four 100th ARW aircraft parked on the ramp at RAF Mildenhall waiting for their next mission.

Above right: A busy ramp at RAF Mildenhall, with five KC-135 Stratotanker aircraft parked on the dispersal.

Left: Boeing KC-13RR from the 100th ARW refuelling a F-16C/J Fighting Falcon from the 52nd Fighter Wing, Spangdahlem Air Base, Germany.

Opposite below: A pair of F-15Es from the 494th Fighter Squadron, RAF Lakenheath format, with the 100th ARW from RAF Mildenhall over Scotland

Above left: General Dynamics F-16C 96-0080 from 480th Fighter Squadron, Spangdahlem Air Base, Germany, during an air-to-air sortie with the 100th ARW, Mildenhall.

Above right: The RAF Boeing E-3 Sentry ZH104 being refuelled by the 100th ARW Boeing KC-135R Stratotanker from RAF Mildenhall.

Above left: The RAF Boeing E-3 Sentry ZH104 joins up with the Boeing KC-135R Stratotanker from the 100th ARW to be refuelled.

Above right: A F-15E Strike Eagle from the 494th Fighter Squadron from RAF Lakenheath joins for fuel with the 100th ARW from RAF Mildenhall over Scotland.

Also, in October 2017, two KC-135s and their crews returned to RAF Mildenhall following a deployment to Naval Station Rota, Spain, supporting Operation Juniper Micron. The USAF has been deploying to Spain in support of operations in Mali and North Africa since 2013.

A KC-135 and their crews arrive in Athens International Airport, Athens, Greece, in the final week of October. The crew were scheduled to participate in short air-to-air refuelling training missions with the Hellenic air force; the training would consist of dry-in-air-refuelling hook ups, allowing the Hellenic Air Force F-16 C/D Fighting Falcons pilots the opportunity to exercise this capability.

November 2017 saw crews from the 100th ARW deploy to RAF Brize Norton, Oxfordshire, in support of exercise Tartan Flag. Tartan Flag is a RAF exercise focused on deployed mission skills, in preparation for Red Flag 2018 at Nellis Air Force Base, Nevada.

USAF aircrew partnered with RAF A330 Voyager aircrew, assigned to the Number 10 and Number 101 Squadron. Successful missions resulted in the first USAF KC-135 and RAF Voyager cell aerial refuelling, when two RAF Tornado GR4s received fuel off the east coast of Scotland.

USAF Lt-Gen. Richard M. Clark, Third Air Force commander, visited RAF Mildenhall on 6 November 2017 and toured the 100th Logistics Readiness Squadron.

USAF Gen. Tod D. Wolters, USAFE and Air Forces Africa commander visited RAF Mildenhall on 9 November; Gen. Wolters arrived in a CV-22 Osprey from the 352nd Special Operations Wing, after visiting another UK base.

In December 2017, RAF Mildenhall locked down following a disturbance on base. The base was locked down and emergency personnel responded to the situation.

The 100th Security Forces Squadron Military Working Dog section suffered a major loss recently when MWD Gina, a German Shepherd, passed away. A memorial ceremony was held in her honour and Team Mildenhall members gathered to pay their respects to the military working dog, who served the Air Force her entire working life.

In January 2018, a formation consisting of a two RAF Typhoons, two USAF F-15C Eagles assigned to RAF Lakenheath, and a USAF B-52 Stratofortress flew over the North Sea off the coast of England. The Stratofortress is part of a deployment to England from the 5th Bomb Wing at Minot Air Force Base, North Dakota.

In February 2018, four USAF F-16C Fighting Falcons, USAF KC-135 Stratotanker, and four Swedish Air Force JAS 39 Gripens flew together during aerial refuelling training in Swedish airspace. The F-16s, part of the 180th Fighter Wing, Ohio Air National Guard, are operating from Amari Air Base, Estonia, as part of a Theatre Security Package in Europe.

On 27 February 2018, twelve USAF KC-135 Stratotankers assigned to the 100th Air Refuelling Wing taxied down the main runway at RAF Mildenhall; this was to show off force manoeuvre and demonstrated the readiness of the wing to provide global air refuelling support at short notice.

In March 2018, two KC-135 Stratotanker refuelling aircraft and approximately sventy-five support airmen from the 100th Air Refuelling Wing, Mildenhall, temporarily deployed to Zaragoza Air Base, Spain, to continue support of Operation Juniper Micron.

In August 2018, thirteen USAF F-15C/D Eagles from the 48th Fighter Wing, RAF Lakenheath, deployed to Keflavik Air Base, Iceland, to provide airborne surveillance and interception capabilities on behalf of NATO and Iceland, fulfilling an annual US commitment. Four of the F-15s will be used for a constant alert mission, ready to intercept and identify unknown aircraft within minutes of notification, while another nine will conduct valuable training in the Icelandic airspace.

In October 2018, the 100th ARW supported a large forces exercise hosted by RAF Lakenheath, which included refuelling F-22 Raptors from Joint Base Langley-Eustis, VA; F-15E Strike Eagles from RAF Lakenheath; and F/A-18 Super Hornets from the Carrier Air Wing from USS *Harry S. Truman*.

In November 2018, the 100th ARW deployed to Rovaniemi, Finland, in support of NATO-led exercise Trident Juncture, the largest NATO exercise since 2002, with over thirty countries participating. The 100th ARW also deployed to Amari Air Base, Estonia, in support of the 48th Fighter Wing, RAF Lakenheath. During this month, it also worked alongside the RAF F-35 Lightning IIs, USAF F-15E Stike Eagles, and the French air force Dassault Rafale during exercise Point Blank, which was a Nato-led exercise.

In February 2019, Air Force One made a quick fuel stop on its way to the Hanoi and North Korea summit.

With every operational air base, it must have several groups who support each other to achieve the missions and goals it adheres to. RAF Mildenhall, like any other base, is a wing that has distinct missions with specific scopes. It is made up of groups of several squadrons, usually commanded by a colonel. The current base commander, Col. Christopher Amrhein, joined the Air Force in 1996 after earning his commission as a distinguished graduate from the ROTC program at the University of Texas at Austin. He has served in operational and training assignments, logging more than 3,100 hours in a variety of aircraft. He has also supported contingency operations Allied Force, Joint Guardian, Noble Eagle, Enduring Freedom, Iraqi Freedom, and New Dawn

With four different commands, including Air Combat Command, Air Force Special Operations Command, Air Mobility Command, and the USAFE, RAF Mildenhall is a diverse base with a wide range and variety of missions that require a unique form of teamwork—hence why they are known as 'Team Mildenhall'.

The 100th Operations Support Squadron

The 100th Operations Support Squadron provides crucial operation support for the five USAF commands, including mission planning and flight scheduling, tactics and crew communications, and aircrew equipment and weather updates. The squadron also provides support through training on simulators and manages all the wing's flying hour's program. Another vital support the squadron provides is Air Traffic Control, with flight management and data analysis.

The 100ths Maintenance Group

The 100th Maintenance Group is accountable for all the aircraft and its equipment, and its objective is to ensure refuelling operations, tactical airlift, special operations, and strategic readiness are in place. The group provides mission ready aircraft with air refuelling capability in theatre and maintains repairs for all the KC-135 aircraft while able to deploy airmen and equipment in support of NATO and other agencies.

The 100th Aircraft Maintenance Squadron

As part of the 100th Maintenance Group, the 100th Aircraft Maintenance Squadron provides support while performing and managing onsite maintenance to fifteen KC-135R aircraft, including launch, recovery, special inspections, service, troubleshooting, and component replacement. It also supports air refuelling by providing maintenance recovery teams for all

Vital to airfield operations, the view from inside RAF Mildenhall's control tower.

KC-135 aircraft operating within Atlantic, European, and Southwest Asian theatres. The 100th Maintenance Operation Squadron is further responsible for monitoring the overall maintenance production effort through scheduling, analysis, and assessing the health of the fifteen KC-135s.

Finally, the 100th Maintenance Squadron performs all the relevant aircraft, component, and ground maintenance tasks and support for ACC, AFSOC, and AMC, USAFE, AFRICOM, and transient aircraft. The squadron also maintains a combat ready status, which can deploy personnel and equipment in support of NATO, United States European Command (USEUCOM), and Unites States Special Operations Command (USSOCOM) operations throughout Europe and Africa.

The 100th Mission Support Group

The 100th Mission Support Group consist of five squadrons that provide support for the United States Air Force Major Air Command (MAJCOM) flying program and the USAFE air refuelling programme across Europe and Africa. The group maintains RAF Mildenhall as an air base with the operating support quality of life facilities, material, and services to approximately 16,800 military, civilian, dependent, and retirees. This also includes personnel working in the UK, 501st Combat Wing, the US Navy, and the US Army, plus over thirty other locations.

The 100th Civil Engineer Squadron

The 100th Civil Engineer Squadron provides RAF Mildenhall with a comprehensive civil engineering department, with over 12,000 personnel from ACC, AFSOC, USAFE, and the US Navy, making it a very diverse squadron.

Ground crew working on a sheet metal piece that will be replaced on a Boeing KC-135R

The squadron constructs, maintains, and repairs almost anything on the base, from property (which could be any form of maintenance or repair), environmental (which could include pollution prevention, hazardous waste, recycling of items, and the conditions of the roads on base), to fire protection and emergency management (which again could be a full-scale aircraft emergency to an oven fire in a building).

The 100th Civil Engineer Squadron Fire Prevention Flight

100th Civil Engineer Squadron Fire Prevention Flight deals with all aspect of fire prevention within and outside of RAF Mildenhall, up to a 15-mile radius. With a response time of less than five minutes, the fire department has an amazing track record; this is due to the continued training with major drills twice quarterly. The fire department works very hard together to achieve their mission objective, and in 2016, it received the 'Fire Department of the Year' award.

The Civil Engineer Squadron is also responsible for the physical aspects of the base, which includes over 1,600 acres, 1,500 facilities (of which 110 is housing), and over 9,000 feet of runway. The squadron also maintains over fifty-five mobility teams, which have a total of 135 personnel providing a twenty-four-hour response to emergencies system. The squadron is very impressive, as 'Team Mildenhall' has shown through their professionalism. During a single day, a civil engineer may unblock toilets, fix street lights, retarmac part of the runway, and then put a fire out.

Three fire tenders sit peacefully at RAF Mildenhall's fire station.

Above: Fire tender 12 parked in the station with the fire crews' protective clothing on the floor, prepared for the next call out.

Below: Two fire tenders parked in the fire station, prepared for the next call.

A mock-up of an aircraft made of steel to re-enact fire scenarios. The aircraft has internal and external parts, such as the fuselage, undercarriage, and engines.

A training building that allows firefighters to re-enact any fire scenarios.

The 100th Communications Squadron

The 100th Communications Squadron operates and maintains a voice and data network, plus installed air traffic control landing systems, transmission, and satellite systems for the 100th ARW, USAFE-UK, ACC, AMC, and AFSOC partner units. The squadron also provides a full postal operation and information service for more than 12,000 personnel and their dependants. The squadron also protects the base from cyber intrusion and attacks.

The 100th Force Support Squadron

The 100th Force Support Squadron provides support to the five MAJCOM and the combat forces assigned to the four combat commands. The squadron also provide support, morale boosting, and welfare to over 16,000 personnel and their dependants by enhancing the quality of life through personnel services, including education, recreation, and mortuary programs.

The 100th Logistics Readiness Squadron

The 100th Logistics readiness Squadron is a very large department broken up in sections, or flights, which provides comprehensive logistic support to the 100th ARW and the other units at RAF Mildenhall. The squadron stores and distributes supply assets and provides passenger travel, cargo movement, and household goods shipments to other US air bases globally. The squadron is responsible for all transportation of cargo from household goods to war reserve materiel. On occasions, third party companies are used for shipment assignments, and this will be coordinated by them. The squadron also provides and supports the base with

The 100th Logistics readiness Squadron (LRS) has a vast array of vehicles that service RAF Mildenhall.

transportation for movement of aircrew, cargo, and refuelling trucks, with over 550 vehicles, and it has issued over 25 million gallons of aircraft fuel. Once again, respect is warranted to 'Team Mildenhall' for its teamwork and mission accomplishment.

The material management flight (LGRM) is responsible for stocking, storing, managing, and inspecting department of defence (DOD) supplies and equipment. The LGRM primary role is to liaise between customers and the USAF Global Logistics Support Centre (AFGLSC).

Asset management is responsible for the storage of aircraft parts (APS), hazardous materials and items (HAZMAT), individual equipment element (IEE), and individual protective equipment (IPE). The department is also responsible for inspecting all the above items.

The customer support department has two sections: customer support liaison and equipment accountability. Customer support manages materials and liaises with third party distributers while the equipment accountability deals directly with the wing.

The Maintenance Support Department

The Maintenance Support department consists of two sections: the maintenance support liaison (MSL) and flight service centre (FSC). MSL is responsible for the overall maintenance and material problems plus the review and coordination of personnel training when required. FSC provides support to the maintenance support section by controlling all the items that require replacing, this will be from the time the item becomes unserviceable until it is removed and then replaced.

Deployment and distribution flight (LGRD) is responsible for planning and administration of all the wings operational deployments and the distribution of air force cargo, passengers, and private property. The LGRD is also responsible for the management and distribution of the wings war reserve materiel (WRM). The distribution flight is responsible for the logistical movement of cargo and personnel on a daily operation. This section monitors receipts of shipments for the department of defence supplies, vehicles, cargo, and equipment.

The Small Air Terminal and Passenger Movement section provide official travel services for all Department of Defense passengers. This can consist of either passenger service flying via USAF aircraft, which can be seen in and out of Mildenhall regularly as they transit to and from their destinations, or commercial travel, which would be a third-party airline. With the use of the Small Air Terminal, RAF Mildenhall can and has been used for VIP movements when civilian airfields security arrangements cannot meet the demands required for the VIP.

Fuels Management Flight

Fuels Management Flight is split into three sections: operation, quality assurance, and fuels information service centres. Operations are responsible for refuelling all aircraft in a timely manner; fuel is pumped from a defence fuel supply point in Thetford via pipeline then transferred into a bulk fuel storage station. When transient aircraft arrive, they would have informed the fuels information and service centre on the amount of fuel required for their onward journey. It is operations responsibility to fuel the aircraft with the correct amount of fuel, and this is done via an underground pipeline that goes directly to every aircraft hardstand at RAF Mildenhall. The fuel hydrants are connected to the underground pipeline, which is connected to the fuel bulk storage tanks that hold millions of gallons of fuel, or, alternatively, technicians can deliver over 6,000 gallons of fuel via the fuel bowers (trucks).

The material management flight (LGRM) warehouse, where ground personnel sort all the incoming and outgoing mail for global transitioning

The material management flight (LGRM) warehouse, cargo, and normal mail checking in area that will be shipped to other US bases.

Used gas masks that need replenishing and cleaning before personnel can reuse.

Presidential Bell MV-22B Osprey 168284 parked on the dispersal at RAF Mildenhall as the summer sun begins to set.

KC-10 Extender 87-0123 from Travis Air Force Base departing RAF Mildenhall heading back to the States.

Above: Boeing C-32A transport the US leaders and VIPs around the globe, seen here arriving in the late evening at RAF Mildenhall.

Below: Learjet Model C-21A from the 86th Airlift Wing, Ramstein, Germany, arriving at RAF Mildenhall.

Sikorsky VH-3D Sea King 159355 lifting off from the ramp at RAF Mildenhall during President Obama's visit to the UK in 2011.

Above: The Boeing E-3B Sentry from the 18th Wing on finals for runway 11 at RAF Mildenhall.

Below: The Rockwell B-1 Lancer 85-0064 parked on the ramp at RAF Mildenhall after diverting due to engine issues.

US C-17 transporter arriving at RAF Mildenhall in a glorious sunrise.

US C-130 Hercules arriving at RAF Mildenhall during a beautiful sunset.

Vital fuel pipeline that supplies all the aircraft that visit RAF Mildenhall.

Ground personnel travelling to the next aircraft that requires fuel.

The quality assurance section is responsible for making sure all departments within the flight are reaching their mission standards in accordance with established standards and procedures.

The fuels information service centre receives calls from dispatch operators and customers requiring fuel. The normal response time from a call being received to an aircraft having been assigned fuel trucks and technicians is fifteen minutes—this could be for either visiting or based aircraft. The fuels information service centre overseas the fuel labs, which are constantly checking fuel for any form of contamination that could cause aircraft damage, and tests are completed to ensure there is no water or solid particles in the fuel.

The 100th Security Forces Squadron

The 100th Security Forces Squadron protects and defends the 1,600-acre boundary of RAF Mildenhall and its assets. The squadron provides law enforcement all year round, protecting the 16,500 military and civil personal and their dependents based at RAF Mildenhall. The squadron support and maintain base security by working alongside the MOD at the main gates for visitors and allowing USAF personnel to gain entry to the base. The squadron also maintains law and order on the base, which includes basic base driving laws, personnel domestic issues, and base intrusion from third party threats.

The squadron also maintains over thirty mobility ready combat teams ready for deployment at any notice. The squadron can be seen along the perimeter fences at RAF Mildenhall, administering security checks either in trucks or military dog walking. It has a mutual understanding with the UK's aviation enthusiast community and their love for the hobby of watching aircraft: the aviation enthusiasts are known as a form of information gathering and protection, which is welcomed by both RAF Mildenhall and RAF Lakenheath.

The Personal Financial Readiness Program

The Personal Financial Readiness Program at the Airman and Family Readiness Centre, located at Washington Square, offers individual financial planning assistance to any DOD ID card holder. The department deals with all airman financial aspects while on tour in UK.

US personnel find the cost of living in the UK more expensive than living in the US, and personnel are advised to start saving for the move months before their deployment to the UK. To help offset the higher price of non-housing goods and services, personnel receive an Overseas Cost of Living Allowance (COLA).

When personnel make the transition to the UK, they will encounter start-up costs; this transition may require several set up costs, including new appropriate clothing for the climate change (colder), new vehicles, deposits, and the first month's rent for housing, deposits for utilities, British household items, etc. A lot of personnel also have a loss of income: if they are married, they must take into consideration the possible loss of a second income either for the duration of the tour or a small time while their spouse is job searching. Many American financial institutions will not allow new bank and investment accounts to be opened while living in the UK due to regulatory agreements.

Financial considerations for personnel range from vehicles to household items. The UK requires a yearly MOT car inspection, US vehicles shipped over would cost several hundred dollars, but fuel is still charged at US prices.

Security forces apprehend an intruder close to the runway at RAF Mildenhall.

RAF Mildenhall security forces K-9 section conducting training.

With home rental, typically in the UK, personnel must pay the first month's rent (which can range from £700 to £2,500) and a security deposit equivalent to a month's rent in advance. Personnel are advised to get a house under their Overseas Housing Allowance found online, otherwise, as the pound–dollar exchange rate fluctuates, this may have a significant impact on their out-of-pocket expenses. In the UK, VAT is like sales tax in the US. VAT is a British tax (20 per cent) on all products and services except for children's clothing and food. This tax is not applicable to items purchased on base.

The 95th Reconnaissance Squadron

The 95th Reconnaissance Squadron is assigned to the 55th Operation Group Air Combat Command, stationed at Offutt Air Force Base, Nebraska, and the squadron houses several variants of the Boeing RC-135 aircraft, equipped for reconnaissance duties.

The 95th Reconnaissance Squadron is one of the oldest units in the USAF, being initially formed as the 95th Aero Squadron in 1917 at Kelly Field, Texas. The squadron fought on the Western Front during the First World War.

During the Second World War, the unit was deployed to the Mediterranean as part of the 12th Air Force as a B-26 light bomber squadron participating in North Africa and South France.

During the Cold War, the squadron was involved in the Korean War, equipped with the B-26 Invader bomber and then finally flying the TR-1A Dragon reconnaissance aircraft while supporting NATO. After the end of the Cold War, the 95th RS was no longer needed, and the unit was deactivated in 1993. This did not last long as the unit was reactivated in 1994 at RAF Mildenhall, this time flying RC-135 Rivet Joints, OC-135 Open Skies, and E-4B aircraft.

The 95th RS is equipped with RC-135 Rivet Joints, conducting flight operations in the European and Mediterranean theatres. It provides all operational management, aircraft maintenance, administration, and intelligence support to provide politically sensitive data vital to national foreign policy.

As a component of the 55th Operations Group, the squadron's main flying operations are conducted from RAF Mildenhall, UK, and Souda Bay, Crete. The squadron will be relocating to RAF Fairfield in 2023.

Boeing OC-135B Open Skies 61-2672 on the underrun runway 11, departing RAF Mildenhall, August 2016.

Boeing RC-135U 64-14847 departs RAF Mildenhall on an early morning mission, expected to last at least twelve hours.

Boeing NC-135W 61-2666 lifts off from RAF Mildenhall, heading back to Offutt Air Force Base, Nebraska.

Above: Boeing RC-135U 64-14847 arriving back to RAF Mildenhall after a long mission to an undisclosed destination.

Below: Boeing RC-135W 62-4125 at the holding point for runway 11 at RAF Mildenhall before departing towards the Middle East.

Boeing RC-135W 62-4125 departing runway 11 at RAF Mildenhall, heading towards the Middle East on a long-extended mission.

Boeing RC-135W 62-4138 awaits departure clearance from ATC at RAF Mildenhall.

352nd Special Operation Wing

The 352nd Special Operations Wing (SOW), based at RAF Mildenhall, is a crucial part of USAF Special Operations Command. The wing has more than 1,000 air force personnel assigned, plus over forty personnel attached from Joint Special Operations Air Component-Europe assigned to the 25th Intelligence Squadron. The 352nd SOW is the only air force special operations unit in European Command.

The 352nd is under the operational control of Special Operations Command Europe and performs specialised operations using advanced aircraft, tactics, and air refuelling techniques to transport and resupply military forces globally. The 352nd SOW provides tilt-rotor aircraft aerial refuelling and special operations weather capabilities. The unit has six squadrons and two distinct types of aircraft: the MC-130J Commando II and the CV-22B Osprey, the USAF's premiere tiltrotor aircraft that arrived at RAF Mildenhall in June 2013.

The 352nd has deployed in a vast array of support operations, including Desert Storm, Provide Comfort, Enduring Freedom, Iraqi Freedom, Provide Promise, Deny Flight, Allied Force, Assured Response, Silver Anvil, Atlas Response, Support Hope, Joint Endeavor, Deliberate Force, Guardian Assistance, Shining Express, Autumn Return, Assured Delivery, Aster Silver, Enduring Freedom-Trans Sahara, Odyssey Dawn, and Unified Protector. The wing regularly supports special operations overseas.

The 352nd Special Operations Support Squadron

The 352nd Special Operations Support Squadron provide in-garrison and deployed command and control and operational support for the two flying squadrons, one special tactics squadron, and one maintenance squadron. The support includes logistics and contingency planning, aircrew training, communications, intelligence, and security force protection.

The 352nd Special Operations Maintenance Group

The 352nd Special Operations Maintenance Group provides all operational maintenance on the MC-130J Commando II and the CV-22B Osprey aircraft assigned to the two special operations flying squadrons.

Five Lockheed Martin C-130J Hercules parked on the 352nd dispersal at RAF Mildenhall.

Three Bell Boeing V-22 Ospreys parked on the 352nd dispersal at RAF Mildenhall.

The 7th Special Operations Squadron

The 7th Special Operations Squadron operate the CV-22B Osprey, flying night, adverse weather, long-range troop transport and resupply operations into potentially hazardous areas. The squadron also supports non-combat evacuations, humanitarian relief, and other operations.

The squadron is a component of the 753nd Special Operations Group (SOG), the US Special Operations Command based at RAF Mildenhall. The squadron can be deployed to extract troops from hostile or other undesirable locations. In 1964, the 7th Special Operations Squadron was activated as the 7th Air Commando Squadron (ACS/USAFE) at Sembach Air Base, West Germany. The new squadron was equipped with four C-123s, six C-47s, and two U-10 aircraft, and had 265 authorised personnel. In 1967, the 7th ACS was designated as the unit to receive the new C-130E Combat Arrow aircraft and command of the squadron was transferred from USAFE to Seventeenth Air Force.

In 1968, the squadron's move to Ramstein Air Base in West Germany also saw the C-123s and U-10s replaced by the MC-130 Aircraft, and the squadron was renamed the 7th Special Operations Squadron (SOS). The squadron continued their training on the MC-130E until the spring of 1971, when the squadron was relocated to Rhein-Main Air Base, near Frankfurt. By 1973, only four MC-130s remained at the air base, with all the others either decommissioned or returned to America. The 7th SOS's MC-130Es, codenamed 'Combat Talon', have been spotted everywhere in Europe, and most sightings would, perhaps, have been connected to a NATO-led marine unit exercise that involved them.

In 1991, the unit deployed to Incirlik AB, Turkey, in support of Operations Desert Shield/Storm. Within twenty-four hours of the President's order, the unit performed the first operational mission: the Operation Provide Comfort airdrop over northern Iraq. Flying operations continued until the end of the war. In the later part of 1991, the squadron was notified that it would move from Rhein-Main to RAF Alconbury, but was subsequently deployed to support the evacuation of American citizens in Sierra Leone. Operation Silver Anvil successfully brought home more than 400 people.

During the move to RAF Alconbury in 1992, the unit converted from the MC-130E Combat Talon I to the MC-130H Combat Talon II. During 1992–1994, the 7th Squadron was deployed worldwide, including on a humanitarian relief mission to Rivne, Ukraine, where there was an urgent need for supplies for people in Morocco. The squadron also deployed training detachments to Greece and low-level night routes in Spain. The squadron's stay at RAF Alconbury was a very short one, and in January 1995, the squadron moved to RAF Mildenhall, where the 352nd Special Operations Group consolidated all its assigned forces.

Today, the 7th SOS continues to maintain its traditional ties with the UK, Norway, Germany, Spain, Italy, and France. From Central Asia to South Africa and all of Europe, the squadron's mission is ever expanding, and it is ready, willing, and able to perform wherever tasked—anywhere, anytime. In 2013, the 7th SOS converted from the MC-130H to the Bell Boeing V-22 Osprey, and on January 2015, the last MC-130H departed RAF Mildenhall, returning to Hurlburt Field where they will continue to fly and make an impact for special operations.

The 67th Special Operations Squadron

The 67th Special Operations Squadron recently completed transitioning from the Lockheed MC-130P Combat Shadow in 2014 to the MC-130J Commando II Aircraft, which are now

Bell Boeing V-22 Osprey on short finals for runway 11 at RAF Mildenhall.

Above: Bell Boeing CV-22B Osprey 09-0046 departs RAF Mildenhall on a general handing exercise.

Below: Bell Boeing CV-22B Osprey 11-0060 await air traffic control permission to depart RAF Mildenhall on a navigational training sortie.

Bell Boeing CV-22B Osprey 11-0057 arrives back to RAF Mildenhall after a short training exercise to Holbeach Range.

based at RAF Mildenhall. The MC-130J Commando II provide precise, reliable, flexible, and responsive specialised air mobility. The aircraft penetrate possibly dangerous areas to conduct single- or multi-ship invasion, evacuation, and resupply of special operations forces via airdrop or air land operations and conduct long-range refuelling operations. The squadron also supports information operations, humanitarian relief, medical evacuations, and non-combat evacuations.

The 67th Air Rescue Squadron

The 67th Air Rescue Squadron was activated in November 1952 at RAF Sculthorpe, then discontinued and deactivated in March 1960 at Prestwick, Scotland. It was activated again in 1961, and the unit was redesignated the 67th Air Recovery Squadron in 1965, and then as the 67th Aerospace Rescue and Recovery Squadron in 1966.

In the middle of 1966, it was relocated to Morón Air Base, Spain, but returned the UK and RAF Woodbridge in 1970. The unit operated the Lockheed HC-130 fixed wing; it also used rotational support for its detachment of H-3 helicopters stationed at Keflavik, Iceland, and Sikorsky MH-53 rotary wing aircraft.

In 1988, the unit was split into two units, redesignated the 67th Special Operations Squadron for the HC-130 aircraft and the 21st Special Operations Squadron for the HH-53 rotary wing.

The 321st Special Tactics Squadron

The 321st Special Tactics Squadron provides a fast reaction, rapid, deployable force capable of establishing and providing control of the air-to-ground interface during special operations missions. Unit combat controllers conduct the reconnaissance, surveillance, and establishment of assault zone sites and provide air traffic control and long-range secure command and control communications. The unit also provides combat trauma medical care, personnel recovery, and terminal attack control of munitions delivered by fixed- and rotary-wing aircraft. The squadron also has special operations weathermen assigned to provide weather support for other US and NATO military forces.

Lockheed Martin C-130H 87-0023 flying overhead before arriving at RAF Mildenhall.

Above: Lockheed Martin C-130P 65-0991 arriving back at RAF Mildenhall during the late afternoon.

Below: Lockheed Martin C-130J 12-5759 holds while a base KC-135R Stratotanker departs RAF Mildenhall on a training mission.

4
RAF Mildenhall-Based Aircraft

The Boeing KC-135 Stratotanker

The Boeing KC-135 Stratotanker is a military aerial refuelling aircraft, replacing the KC-97 Stratofreighter. The KC-135 was the USAF's first jet-powered refuelling tanker. It entered service with the USAF in 1957 and was initially tasked to refuel strategic bombers but was then used in the Vietnam War and later conflicts, such as Operation Desert Storm in Iraq, to extend the range and endurance of US tactical fighters and bombers.

The KC-135 has over fifty years of continuous service and many could be flown until at least 2040. It is supplemented by the larger KC-10 and will ultimately be replaced by the Boeing KC-46 Pegasus.

The KC-135 looks like the Boeing 707 airliner but has a narrower and shorter fuselage and is structurally different to the airliner. In 1954, the USAF placed an order for over twenty-five KC-135A, and, with a final production figure of 820, these were the first of many in the KC-135s family.

The first KC-135 flew in the summer of 1956, with the first operational aircraft delivered in 1957. Amazingly, the last was not delivered to the air force until 1965. The KC-135 had many variants during its service, which included the RC-135 Rivet Joint, which started operations in 1963, plus several KC-135A and KC-135B aircraft that had been modified to EC-135 and OC-135 configurations for use in several distinct roles.

The KC-135Q model was upgraded so it could be equipped to carry JP-7 fuel for the Lockheed SR-71 Blackbird aircraft. The KC-135Q also received the modified CFM-56 engines and was redesignated as the KC-135T and can separate the main body tanks from the wing tanks where the KC-135 draws its engine fuel. The only external difference between a KC-135R and a KC-135T is the presence of a clear window on the underside of the empennage of the KC-135T, where a remote-controlled searchlight is mounted. It also has two ground refuelling ports located in each rear wheel well, so ground crews can fuel both the body tanks and wing tanks separately.

The KC-135R is primarily used for force extension and special operations missions. It is crewed by highly qualified receiver-capable crews. If not used for the receiver mission, these aircraft can be flown just like any other KC-135R.

Boeing KC-135R Stratotanker breaks out of the clouds on an early morning arrival back to RAF Mildenhall.

Above: Boeing KC-135R Stratotanker 59-1492 breaks to the left for a downwind leg before landing runway 29 at RAF Mildenhall.

Below: The Boeing KC-135R Stratotanker lifting off from the runway 29 at RAF Mildenhall on a training sortie.

McDonnell Douglas KC-10 Extender on final approach to RAF Mildenhall.

Above: KC-135R 62-3565 from the 100th ARW, RAF Mildenhall, undergoing maintenance while parked on the dispersal.

Below: Two Boeing KC-135R Stratotankers sits on the dispersal at RAF Mildenhall awaiting their next mission.

The flight deck of the KC-135 Stratotanker.

Boeing KC-135R Stratotanker's engine compartment covers opened during routine maintenance.

During the decades, the KC-135 has undergone numerous upgrades to help expand its capabilities and improve reliability. After many engine upgrades from Pratt and Whitney engines, over 500 KC-135s were modified with the new CFM International CFM56 engines. These high bypass turbofan engines were produced by General Electric and Snecma. The CFM-56 engines can produce 22,500 lb of thrust—a near 100 per cent increase compared to the original J57 engines. The KC-135R can offload 50 per cent more fuel and can go on longer sorties.

The KC-135R has four turbofan engines, which power it through take-offs at gross weights of up to 322,500 lb. Nearly all internal fuel can be pumped through the tanker's flying boom, the KC-135's primary fuel transfer method. A special shuttlecock-shaped drogue, attached to and trailing behind the flying boom, may be used to refuel aircraft fitted with probes. A boom operator stationed in the rear of the aircraft controls the boom while lying prone. A cargo deck above the refuelling system can hold a mixed load of passengers and cargo. Depending on fuel storage configuration, the KC-135 can carry up to 83,000 lb of cargo.

By 2005, the KC-135 fleet was flying over double its planned yearly flying hour program to meet airborne refuelling requirements, which has resulted in more usage and additional costs incurred to the planned maintenance program. In 2009, the USAF indicated that all KC-135s would require skin replacement to allow the aircraft to continue in service beyond 2018.

The KC-135s will be replaced, but with a large fleet, this will take some time. The USAF has selected the KC-46, a Boeing 767-based tanker, as its replacement, and this will slowly be introduced into service.

Above left: Four Boeing KC-135R Stratotankers from the 100th ARW parked on the dispersal at RAF Mildenhall.

Above right: Another busy day on the dispersal of the 100th ARW at RAF Mildenhall, with five Boeing KC-135R Stratotankers awaiting their crew.

The MC-130

The MC-130, built by Lockheed, is operated by the Special Operations Command of the USAF. The MC-130 is based on the C-130 Hercules transport aircraft, but for its missions, the MC-130 is used for infiltration, exfiltration, and resupply of special operation forces personnel, plus air refuelling of special operation helicopters and the V-22 Osprey.

The MC-130 has many different variants, which include the MC-13E Combat Talon I, the MC-130H Combat Talon II, the MC-13W Combat Dragon Spear, the MC-130P Combat Shadow, and, most recently, the MC-130J Commando II.

The USAF requested acquisition of at least thirty-seven new production MC-130J aircraft to replace the Combat Talon and Combat Shadows, with the first ten received in 2011. These are now based at RAF Mildenhall. A total of sixty-nine MC-130J aircraft are planned and these will eventually replace all other MC-130 variants.

Combat Talon, the MC-130E, was developed for special operation missions during the Vietnam War and eighteen were built; four were lost through combat stress, and the remaining fourteen continued in service for over forty years after a few modifications. The updated version of the Combat Talon II was developed in the 1980s and then went into service in the 1990s. Twenty-four were built, but, unfortunately, four were lost in operations.

In 2007, the USAF conducted a review and proposed exploring the prospect of replacement aircraft; at that time, it was hoping for an Initial Operating Capability date of 2018. The USAF has stated its desire for thirty-seven MC-130Js to replace its MC-130Es and MC-130Ps, which are forty or more years old.

Combat Talon was originally developed by Lockheed Air Services at Ontario, California, between 1964 and 1967. It was designated Combat Talon later in 1967. Over the next couple of decades, many upgrades and modifications were made, including electronic and infrared (IR) countermeasures suites and AN/APQ-115 navigational radar, which featured terrain following/terrain-avoidance, Doppler, and mapping radar modes to enable it to operate at low altitudes at night and in all weather conditions and avoid known enemy radar and anti-aircraft weapons.

The Combat Talon I designations were consolidated in 1977 as the MC-130 and have remained under that name since. The Combat Talon became the Combat Talon I in 1984 with modifications of twenty-four C-130Hs to Combat Talon II specifications.

Between 1984 and 1994, a major modification and upgrade program modernised the capability and serviceability of the Talon I extending its service life. All fourteen Combat Talon Is are equipped with upgraded navigational radars, an enhanced electronic warfare suite, and provided with new outer wings. By 1995, all Combat Talon Is were equipped with a helicopter-air refuelling pods.

Talon was involved in numerous amounts of deployments during its service, including the seizure of the US Embassy in Tehran, Iran, in 1979, training operations for a rescue mission of the fifty-three hostages that were begun by Talon crews at Kadena Air Base.

At that time, only seven Combat Talons had the in-flight refuelling capability necessary for the mission. All were assigned to the operation—a complex two-night plan was called Eagle Claw. Talon crews using night vision goggles practised blacked-out landings to insert Delta Force operators and US Army Rangers deep into Iran and developed several methods for delivering extra fuel for the RH-53D Sea Stallion helicopters chosen to carry out the rescued hostages.

Above left: Lockheed Martin C-130H 87-0023 arriving back to RAF Mildenhall after a low-level training sortie.

Above right: Lockheed Martin C-130H 88-0195 holds at the threshold for the active runway while another C-130 departs at RAF Mildenhall.

Above: Lockheed Martin C-130H 85-0021 lifts off into the afternoon sun from RAF Mildenhall on another training sortie.

Below: Lockheed Martin C-130H arrives back to RAF Mildenhall as the sun begins to set after a long mission.

The first phase of the rescue mission began and was led by Lt-Col. Robert L. Brenci of the 8th SOS, call sign Dragon 1. The 1st SOS Talons successfully secured the forward operating location in the Iranian Desert, but the helicopter portion of the mission ended in disaster. Although the mission was an embarrassing failure, costing eight lives, seven helicopters, and an EC-130E aircraft in a ground accident, the MC-130s performed nearly flawlessly. Planning of a second rescue attempt, under the project name Honey Badger, began two weeks after the failed raid. Combat Talon participation in Honey Badger amounted largely to tactics development.

The MC-130E Combat Talon deployed to Incirlik Air Base, Turkey, as part of Operation Proven Force. They supported the first joint search and rescue mission over Iraq, attempting to recover the crew of Corvette 03, a downed F-15E Strike Eagle. However, permission from the Turkish government to fly the mission was delayed for twenty-four hours, and the crew was not recovered.

A Combat Talon I was the first aircraft to land at New Orleans International Airport after Hurricane Katrina in August 2005. On 14 July 2006, the 8th SOS flew its last Combat Talon I mission and began conversion to the CV-22 Osprey, ending forty-one years of active service for the MC-130E Combat Talon I. Although retired from the regular air force, the MC-130E continued to remain in service with the air force reserve command's 919th Special Operations Wing.

MC-130H Combat Talon II testing began in 1982 and the final Combat Talon II configurations, with significant improvements in design, avionics, and equipment, was ready for production. Twelve aircraft were authorised to be purchased. In 1983, the USAF operations forces were transferred to the military command, another move by USAF to divest itself of its special operations role.

The first Combat Talon II was delivered in 1984, but an earlier decision by the USAF not to equip it with the navigational radar suite of the MC-130E slowed its development for years. The first fully operational MC-130H Combat Talon II was in 1991 and three more were delivered in the same year. By the end of the year, the 8th SOS at Hulburt Air Force Base was equipped with six aircraft. The Combat Talon II is equipped with a stronger airframe and lots of modifications to the rear cargo doors. The electronics were also upgraded, which included the global navigational system, plus a special radar for weather and night vision goggles.

The new upgrades to the Talon II allow the aircraft to fly as low as 250 feet, at faster speeds, and provide the ability for more accurate supply airdrops. In 1995, three MC-130H Combat Talon II from the 7th SOS were deployed on a peacekeeping mission to Sarajevo, Bosnia and Herzegovina, as part of Operation Joint Endeavour.

The first Combat Talon II was sent on combat deployment in 1996 during Operation Assured Response. Additionally, in Liberia, Special Forces were deployed to assist with the evacuation of over 2,000 civilians from the American Embassy. By 2000, the Combat Talon II reached full capability with twenty-four MC-130Hs deployed to four squadrons.

They year 2001 saw four Combat Talon IIs infiltrate a taskforce of 199 Rangers from the 3rd Battalion, 75th Ranger Regiment, 658 miles inside Taliban-held Afghanistan. The force dropped onto Objective Rhino, an unused airfield in Kandahar Province 110 metres south-west of Kandahar, to secure a landing zone as a temporary operating base for Special Forces units conducting raids in the vicinity.

Combat Talon IIs of the 7th SOS, with crews from the 15th and 550th SOSs, flew up to fifteen-hour missions, carrying out airdrops and night landing missions from Incirlik Air Base, Turkey, resupplying Special Forces Operational Detachments in Afghanistan during the opening

Lockheed Martin C-130H taxiing to the threshold runway 29 at RAF Mildenhall for a navigation flight to Wales.

Lockheed Martin C-130H, turning short, finals for runway 11 at RAF Mildenhall while doing pattern work.

Lockheed Martin AC-130U 87-0128 departs RAF Mildenhall on the homeward journey to Hurlburt Field Air Force Base.

Lockheed Martin C-130P 66-0220 arrives back at RAF Mildenhall for the last time before heading home to the states and a new home.

phase of Operation Enduring Freedom in 2001. Operating in mountainous terrain, they innovated an airdrop tactic, replicating maximum-effort landing techniques, to rapidly descend from 10,000 feet to 500 feet to ensure accurate gravity drops after clearing high ridgelines into deep valleys.

The MC-130P Combat Shadow provided air refuelling support to the US Army and the USAF helicopters during Operation Just Case. Four Combat Shadows were deployed to Saudi Arabia for Desert Storm and three others were sent to Batman Air Base for Operation Proven Force.

Since the Gulf War, the MC-130P has been involved in numerous operations, including Northern Watch and Southern Watch (Iraq), Deny Flight (Yugoslavia), Restore Democracy and Uphold Democracy (Haiti), Deliberate Force and Joint Endeavour (Bosnia), Assured Response (Liberia), Guardian Retrieval (Zaire), Enduring Freedom (Afghanistan), and Iraqi Freedom.

Now with the newer model MC-130J, with an array of modifications and upgrades, the Commando II will fly low-level aerial refuelling missions as well as infiltration, exfiltration, and resupply missions, taking over from its successful predecessors, the Talon I and Talon II.

Lockheed Martin C-130J Super Hercules

The Lockheed Martin C-130J Super Hercules is the new four-engined turboprop military transport aircraft that has been upgraded from the Lockheed C-130 Hercules.

The Hercules, with its fifty years or more of service, has the longest continuous production run of any military aircraft in history. The new C-130J, with its new engines, flight deck, and navigational system, already has fifteen nations placing orders, with a combined total of 300 aircraft requested, of which over 250 have been delivered.

The C-130J is the newest version of the Hercules and the only model still in production. Externally, it looks like the classic Hercules in general appearance, but the J-model features updated technology. These differences include new Rolls-Royce AE 2100 D3 turboprops with Dowty composite scimitar propellers and digital avionics, including head-up displays (HUDs) for each pilot. These changes have improved performance over its C-130E/H predecessors, such as 40 per cent greater range, 21 per cent higher maximum speed, and 41 per cent shorter take-off distance.

As a cargo aircraft, the C-130J's crew includes two pilots and one loadmaster. No navigator or flight engineers are required unlike the specialised USAF variants—AC-130J, EC-130J, MC-130J, HC-130J, and WC-130J—which may have larger crews, such as navigators and combat systems officers or other specialised officers and enlisted aircrew.

The C-130J variant will replace the ageing HC-130s and MC-130s, and the MC-130J Combat Shadow II aircraft is in production for USAF Special Operations Command, while the RAF has ordered twenty-five aircraft, replacing their C-130 C4 and C5 variants.

The C-130J has been used extensively by the USAF and US Marine Corps in Iraq and Afghanistan. Canada also deployed its CC-130J aircraft to Afghanistan. The C-130Js from several countries have been deployed in support of the US Operation Odyssey Dawn and as NATO's Operation Unified Protector during the 2011 Libyan civil war.

By April 2013, the C-130J surpassed 1 million flight hours, and later that same year, the Indian Air Force performed the highest landing of a C-130J at the Daulat Beg Oldi airstrip in Ladakh at the height of 16,614 feet.

Lockheed Martin MC-130J 12-5760 parked on the dispersal at RAF Mildenhall awaiting its crew before heading out on its next sortie.

Above: Lockheed Martin MC-130J 12-5759 touching down on runway 11 during pattern work at RAF Mildenhall.

Below: Lockheed Martin MC-130J 12-5759 on finals for runway 11 during pattern work at RAF Mildenhall.

Current operators of the C-130J are the USAF, US Marine Corps, US Coast Guard, RAF, Indian Air Force, RCAF, RAAF, Royal Danish Air Force, Royal Norwegian Air Force, Israeli Air Force, and the Italian Air Force.

The Boeing RC-135

The Boeing RC-135 is a large fleet of reconnaissance aircraft built by Boeing and modified by several companies, including Lockheed Martin and General Dynamics. This has been predominately flown by the USAF and, most recently, the RAF to support national level intelligence. The RC-135 can, on occasion, be seen at RAF Mildenhall as this is the aircraft's secondary base, and due to its long-extended missions, it normally departs in the hours of darkness and returns in the hours of darkness.

Based on the C-135 Stratolifter airframe, there are several types of RC-135s that have been in service since 1961, with many variants having been modified numerous times, resulting in an extensive variety of configurations.

The RC-135A was the first variant and was ordered back in 1962 by the USAF to replace the Boeing RB-50 Superfortress and only four out of the nine that were planned were built. They were used for photographic and surveying tasks, based at Turner Air Force Base, Georgia, and later at Forbes Air Force Base, Kansas, as part of the 1370th Photographic Mapping Wing. As technology developed, the missions for the RC-135A were soon taken over by satellites.

The RC-135B was never used operationally but was built for electronic intelligence to replace the RB-47H Stratojet and these were delivered in 1965. Again, modifications to the airframe took place to improve its electronics, and by 1967, with the new modifications, it became the RC-135C, the aircraft being fitted with sideways looking airborne radar (SLAR) antenna on the lower forward fuselage.

The RC-135C was used for strategic reconnaissance duties by the USAF and was equipped with the AN/ASD-1 electronic intelligence system (ELINT). This system gave this aircraft its very own character by the large 'cheek' pods on the forward fuselage, as well as numerous other antennae and a camera position in the refuelling pod area of the fuselage. The RC-135 aircraft are crewed by two pilots, two navigators, and numerous intelligence-gathering specialists. When the RC-135C are fully deployed, all ten aircraft continue in active service as either RC-135V Rivet Joint or RC-135U airframes.

The RC-135D was delivered to Eielson Air Force Base in Alaska in 1962 as part of the 'office boy' project and began operational missions in 1963. The main missions saw the 'Rivets' fly over the northern borders of the Soviet Union and often on missions between Eielson and RAF Upper Heyford and later RAF Mildenhall. However, by the 1970s, and due to the expansion of the RC-135 fleet, the RC135D airframes were converted into tankers.

The RC-135E project name Lisa Ann 'Rivet Amber' operated out of Shemya Air Force Station in Alaska for three years in 1966, and there was only one aircraft built. This airframe was built especially, but the radiation generated by the radar was considered a health hazard to the crew, and both ends of the radar compartments had to be shielded by thick lead bulkheads. This was designed to prevent the forward and rear crew areas from direct contact after boarding the aircraft. The system used could track an object the size of a football from a distance of 300 miles and the missions were to monitor Soviet ballistic missile testing in the re-entry phase. In 1969, Rivet Amber was unfortunately lost at sea and no trace of the aircraft or its crew was ever found.

The next version was the RC-135M, which operated from Kadena Air Force base in Japan during the Vietnam War. Its primary missions were to gather intelligence over the Gulf of Tonkin and Laos. Six RC-135M aircraft were built, and later modifications allowed the aircraft to continue flying in active service as the RC-135W.

Three Cobra Ball aircraft or the RC-135S were built and are part of the 55th Wing, 45th Reconnaissance Squadron, based at Offutt Air Force Base, Nebraska, and were originally assigned to Shemya and used to observe and gain intelligence about ballistic missile tests on the Kamchatka peninsula.

The RC-135T, known as Rivet Dandy, was to supplement the RC-135C/D/M fleet while they were undergoing upgrades. It operated variously with the 376th Strategic Wing at Kadena AB, Okinawa, and the 305th AREFW at Grissom AFB, Indiana, and the 6th Strategic Wing at Eielson AFB, Alaska.

Another variant, the RC-135U Combat Sent was designed to collect technical intelligence on adversary radar emitter systems. The RC-135U data is collected to develop new or upgraded radar warning receivers, radar jammers, and anti-radiation missiles based at Offutt Air Force Base, Nebraska.

The RC-135V/W is the USAF main airborne platform. The RC-135V/W radar systems allows the crew to detect, identify, and geo-locate signals. The mission crew can then forward the gathered information in a variety of formats via Rivet Joint's extensive communications systems. The aircrew consists of the cockpit crew, electronic warfare officers, intelligence operators, and airborne systems maintenance personnel.

Today, all RC-135s are assigned to Air Combat Command, and all are permanently based at Offutt Air Force Base, Nebraska, and operated by the 55th Wing, using various forward deployment locations worldwide, including RAF Mildenhall. The RAF felt they needed to have a replacement for the Nimrod and the RC-135 with its successful program was the perfect airframe.

The first RC-135W arrived at RAF Waddington in 2013 and forms No. 51 Squadron RAF, based at RAF Waddington, along with the RAF's other ISTAR assets, the E-3 Sentry and the Sentinel. They are expected to remain in British service until 2045.

The Bell Boeing CV-22 Osprey

The Bell Boeing CV-22 Osprey is an American multi-mission, tiltrotor military aircraft that has the ability for both vertical take-off and landing (VTOL), and short take-off and landing (STOL).

The V-22 originally came from the US Department of Defense joint-service vertical take-off/landing experimental (JVX) aircraft program, which started in 1981, and the teams from both Bell and Boeing Helicopters were awarded a development contract in 1983 for the tiltrotor aircraft. The first V-22 flew in 1989.

The US Marine Corps began crew training for the V-22 in 2000 and started operations in 2007. These were to replace the Boeing CH-46 Sea Knights. The USAF started operating their version of the V-22 in 2009. Since entering service in both the US Marines and the USAF, it has been deployed to Iraq, Afghanistan, and Libya.

Boeing RC-135V 64-14842 lined up on runway 29 at RAF Mildenhall awaiting air traffic control clearance.

Above: Boeing RC-135U 64-14847 departs RAF Mildenhall early morning on another long mission to an undisclosed destination.

Below: Boeing RC-135W 62-4125 arriving at RAF Mildenhall in the late afternoon from the US.

Above: Bell Boeing CV-22B Osprey departing RAF Mildenhall for a general handling flight to the south-west of the UK.

Below: Call sign 'Knife 01' departs RAF Mildenhall on a training sortie.

The development of the V-22 began back in 1980 after a failed hostage mission where it was demonstrated to the US that the military needed a new aircraft that could take-off and land vertically, but also could carry troops and do so at high speeds. In 1981, the Department of Defense began the joint service program.

In January 1985, the JVX was designated V-22 Osprey, and by March, the first six prototypes were produced. The workloads had been split between Bell and Boeing, with Bell manufacturing the wing, nacelles, rotors, drive system, tail surface, and aft ramp and Boeing manufacturing the fuselage, cockpit, avionics, and flight control. Rolls-Royce integrated the engines.

The US Marine Corps variant of the Osprey received the MV-22 designation and the USAF model received the CV-22. After two of the prototypes of the MV-22 had crashed in 1991–1992, Bell and Boeing redesigned the V-22 to resolved safety issues and made any improvements that were needed. This helped lower production cost, and with this, they redesignated the V-22B model and flights resumed in 1993.

In 1997, a flight test of the four full-scale V-22s began at the Naval Air Warfare Test Centre in Maryland. Unfortunately, in 2000, there were two further fatal crashes of the V-22, killing nineteen marines in total. Once again, the aircraft was grounded while the cause of the crash was investigated. In 2012, many changes had been made to the V-22s hardware, software, and procedures.

In September 2005, the Pentagon approved full production for V-22: forty-eight a year by 2012 and 458 in total. The planned allocation is 360 for the US Marine Corps, forty-eight for the US Navy, and a further fifty for the USAF at an average cost of $110 million per aircraft.

In 2013, the US was reportedly hoping to sell over 100 V-22 internationally, with up to fifteen countries interested—these included Israel, the United Arab Emirates, and Japan. Also, in 2013, Bell began to lay off their workers on the V-22 production line following the implementation of defence cuts.

The Osprey is the world's first tiltrotor aircraft with one three-bladed prop rotor, turboprop engine, and transmission nacelle mounted on each wingtip. It operates as a helicopter, and once airborne, the nacelles rotate forward 90 degrees in as little as fifteen seconds for horizontal flight, making the V-22 a more fuel efficient, high-speed turboprop aircraft. The V-22's rotors fold in ninety seconds and its wing rotates to align, front-to-back, with the fuselage. To stop potential damage caused from heat of the V-22's engines, the Naval Air Systems Command devised a temporary fix of portable heat shields placed under each engine. However, a long-term solution would be required and a redesign of the carrier decks with heat-resistant coating was acquired to solve this problem.

The Osprey is equipped with a glass cockpit that incorporates four multi-function displays compatible with night vision goggles, a shared central display unit that displays various images including Digimaps, imagery plus primary flight instruments, navigation, and landing aids (TACAN, VOR, ILS, GPS, and INS), and system status. The flight director panel of the cockpit management system allows for fully coupled (autopilot) functions that take the aircraft from forward flight into a 50-foot hover with no pilot interaction other than programming the system.

The Osprey carries an Impressive armament load that includes a M240 machine gun or .50-calibre M2 machine gun on the loading ramp, which can be fired rearwards when the ramp is lowered, a .50-calibre GAU-19 three-barrel Gatling gun mounted below the V-22's nose, plus a BAE Systems developed belly-mounted, remotely operated gun turret system named the Interim Defensive Weapon System (IDWS). The IDWS is remotely operated by a gunner inside the aircraft, who acquires targets via a separate pod using colour television and forward-looking infrared imagery. The IDWS was installed on half of the first V-22s deployed to Afghanistan in 2009.

Since entering service, the Osprey has been deployed to many combat zones and the MV-22 was deployed to Afghanistan in November 2009 with VMM-261 known as the 'raging bulls' based at Marine Corps Air Station (MCAS) New River, North Carolina. It saw its first offensive combat mission, Operation Cobra's Anger, on 4 December 2009. Ospreys assisted in inserting 1,000 Marines and 150 Afghan troops into the Now Zad Valley of Helmand Province in southern Afghanistan to disrupt Taliban communication and supply lines. In February 2011, Marine Commandant Gen. James Amos indicated MV-22s deployed to Afghanistan had surpassed 100,000 flight hours and were noted as 'the safest airplane, or close to the safest airplane' in the Marine Corps' inventory.

Bell Boeing CV-22B Osprey parked on the dispersal at RAF Mildenhall.

In 2010, the MV-22 was sent to Haiti as part of Operation Unified Response relief efforts after the earthquake there—the type's first humanitarian mission. In 2011, two MV-22s from Kearsarge participated in a mission to rescue a downed USAF F-15E crew member during Operation Odyssey Dawn, and on 2 May 2011, following Operation Neptune's Spear, the body of Osama bin Laden, founder of the al-Qaeda terrorist group, was flown by a MV-22 to the aircraft carrier *Carl Vinson* in the Northern Arabian Sea, prior to his burial at sea.

In 2014, three MV-22Bs were placed on alert at Al Jaber Air Base in Kuwait to be ready within thirty minutes to recover downed pilots during the military intervention against ISIL. On twenty-nine occasions between 1 November and 24 April 2015, two Ospreys and a KC-130J aerial tanker assigned to this Tactical Recovery of Aircraft and Personnel mission spent 145 flight hours loitering, ready to perform rescue missions if required. The only pilot that was downed was a Jordanian, but he did not have a radio on him when he ejected and landed too close to ISIL forces. Also, in 2014, V-22 aircraft carried Delta Force commandos to a campsite in eastern Syria, where Islamic State militants had held Americans and others hostage. The commandos quickly eliminated the militants at the site but found that the hostages had been moved elsewhere and returned home empty-handed.

The US Navy's variant will incorporate an extended-range fuel system, a high-frequency radio for over-the-horizon communications, and a public-address system to communicate with passengers and will be designated as the HV-22 with four initially being bought each year between 2018 and 2020.

RAF Mildenhall and 352nd Special Operations Wing have acquired ten CV-22 Ospreys now and fly many special operation missions during the week, normally flown between the late afternoon and evening. These aircraft will be reallocated to Spangdahlem in Germany once the proposed date of RAF Mildenhall closes in 2024.

Above left: Two Bell Boeing CV-22B Osprey parked on the dispersal ramp at RAF Mildenhall.

Above right: Bell Boeing CV-22B 07-0033 lining up for runway 11 at RAF Mildenhall on a grey, murky day.

Bell Boeing CV-22B 11-0057 approaching the threshold runway 11 at RAF Mildenhall during pattern work.

5
A Day in the Life of 100 ARW Air Crew

The average day for the crew can vary by where you are and what mission you are doing. Missions typically demand long hours, normally around eight to twelve hours in the air. During deployment, a typical day would be wake up, go to the gym, fly for eight to ten hours, land, eat, see the family, and then go to sleep—then repeat it a day or two later.

A basic flight crew consist of three—two pilots (better known as aircraft commanders) and a boom operator—but for longer missions, this can be extended to three pilots, and the number of boom operators are depending on the passenger count and how many refuelling's are planned for that flight.

The crew arrive for their mission briefing normally three hours before their departure—this would be four hours if the mission is of high priority. All three crew members will review all aspects of the flight—refuelling, passenger/cargo, or on deployment for operations.

During the briefing, the aircraft commander confirms all aspects of the forthcoming mission to the co-pilot and boomer, who will also present their own section of the brief; the crew must confirm they understand every instruction during the briefing. The briefing covers subjects such as fuelling procedure, fire protection, and ditch procedure if the inevitable happens. Once this briefing is complete, the aircraft commander will start another briefing, which lasts around thirty minutes and is more in-depth. This briefing covers the amount of fuel required for the mission, including the KC-135's fuel as well as the fuel for the receivers (aircraft to be refuelled), the number of receivers, the divert airfield, the on-route waypoints, and the routing the flight will be taking.

The final briefing, which last about twenty minutes, is the weather brief. This is broadcasted via satellite video link, and it covers all aspects of the weather, from departure airfield, weather *en route*, and any divert airfields, plus the long-range forecast for the return journey back to RAF Mildenhall. The KC-135 wind threshold is around 20 knots crosswind—any stronger and the aircraft would have to divert.

With the briefings complete, the crew have a few moments to refresh and use the toilet before transport arrives to take them to their aircraft. Once the crew arrive at the aircraft, the aircraft commander is met by the crew chief; they will discuss any maintenance issues that have been rectified or will not affect the mission. It takes around forty minutes from arriving at the aircraft to actual engine start.

The fuel, passenger, and cargo payload are discussed and confirmed by both parties, the aircraft commander will complete a visual walk around the aircraft, and once the aircraft

Above left: Aircraft Commander Capt. J. Fuller conducts the briefing to his co-pilot and boomers.

Above right: Aircraft Commander Capt. J. Fuller conducts the second briefing, which is more in depth but in a relaxed environment.

Above left: Aircraft Commander Capt. J. Fuller and his co-pilot, Capt. R Barnes, take notes during the weather satellite link up brief.

Above right: Aircrew take a few minutes to relax and talk to each other about their forthcoming sorties.

Both the crew chief and aircrew go through vital paperwork on the ground before boarding the aircraft.

Above left: Capt. R. Barnes conducts visual checks of the KC-135R Stratotanker before the mission.

Above right: Capt. R. Barnes and the crew chief carry out visual checks of the aircraft before it is deemed airworthy.

commander is satisfied, the aircraft is airworthy, and the payload is correct they will board the KC-135. The flight crew will now begin their pre-flight checks, with the pilots reading through every item—disciplined and interacting with each other. With the checks completed, radio transmissions begin with a call to Mildenhall tower and the base command post.

Command post (better known as 'banter ops') is a department within the 100th ARW that holds all the relevant information and is sent to and from aircraft that arrive or depart RAF Mildenhall.

The crew have requested and have been approved for engine start up; after a few minutes, all four engines from the KC-135 are running and another call is made to the tower requesting permission to taxi, which again is approved.

Aircraft Commander: Quid 78, request taxi.
Tower: Quid 78 taxi runway 29 via taxiway Bravo.

Checks continue before the aircraft commander releases the brakes and applies power to move the aircraft from its parking area; the aircraft then slowly taxies towards the active runway while more pre-flight checks are carried out.

Call sign for the flight is Quid 78 with a departure time of 8.20 a.m., arriving back to Mildenhall at 4.50 p.m. The flight was to head north towards air refuelling areas (ARA) 4, which are located around the UK. The aircraft commander calls the tower once again, requesting take-off clearance. This is granted:

Aircraft commander: Mildenhall tower Quid 78 ready for take off
Tower: Quid 78 cleared for take-off runway 29, winds 270 at 5 knots climb and maintain runway heading 2,000ft and contact Lakenheath departures.

Capt. J. Fuller and Capt. R. Barnes begin the pre-flight checks.

Pre-flight checks are carried out between the two crew members using the latest technology.

Boeing KC-135R Stratotanker 59-1471 sitting on the dispersal waiting for engine start-ups.

Above: Boeing KC-135R Stratotanker taxies to runway 29 at RAF Mildenhall, ready for the forthcoming sortie.

Below: Boeing KC-135R Stratotanker lifting off from the runway 29 at RAF Mildenhall.

The power is applied as the aircraft rolls onto runway 29 at RAF Mildenhall. The aircraft commander applies more thrust into the engines and the aircraft begins its take-off roll down the runway.

The KC-135 gets to its rotation speed (approximately 160 knots) and is airborne within 7,000–8,000 feet of runway, dependent on the weight of the aircraft. As the aircraft passes the 2,000-foot altitude, the aircraft commander switches frequency on the radio to communicate with RAPCON at RAF Lakenheath, granting further climb and heading instructions.

Quid 78 now climbs to 19,000 feet and heads direct to waypoint Marham. With further checks being completed by the crew, RAPCON controllers inform the crew to change frequency again to London Military, who will now control them for much of the onward flight.

London Military control give Quid 78 further waypoints and their final climb instruction to 27,000 feet. With that, the crew begin their refuelling checklist and the boom operator moves to the boom pod at the rear of the aircraft and raises to the covers over the windows. Lying down, the boomer manoeuvres the boom in accordance with a checklist to ensure the system is fully operational for the arriving aircraft.

Quid 78 has now reached the ARA4 and begins the circular pattern, awaiting the receivers—US F-15s from the 48th Fighter Wing 494th and 492nd Squadrons, General Dynamics F-16s from 52nd Fighter Wing Spangdahlem Air Base, and Boeing E-3 Sentrys from RAF Waddington. The boomer must now maintain contact with each receiver, communicating directions to the aircraft and boom; fighters must join with the tanker from the left-hand side of the aircraft. The boom is hydraulically moved into place, and while the receivers gain fuel, the aircraft commander will keep the aircraft at the same altitude and keep with the ARA.

If there are multiple receivers, they will remain in formation with the tanker until all the aircraft have been refuelled. Normally, the receiver would call for disconnection from the boom and the receiver would then throttle back and descend until clear of the boom. On completing the refuelling, the formation of receivers will pull away and the tanker will begin its journey back to RAF Mildenhall. The boomer will raise the boom to its stowed position and return the window cover to its place.

On return to Mildenhall, Quid 78 communicates with London Military, who gives the aircraft commander the descent command and heading. The crew now begin their pre-landing and descent checks.

With the aircraft approaching Mildenhall airspace, the aircraft commander is again asked to change frequency back to RAPCON, who will give instructions for the further descent into Mildenhall. The aircraft commander has sufficient fuel remaining for the crew to practice some pattern work (known as 'touch and go' landings) and informs RAPCON of the aircrafts intentions; RAPCON gives the aircraft commander further commands and asks for a frequency change to RAF Mildenhall tower, who communicate with Quid 78 until it is on the ground.

After a couple of hours in the pattern, the aircraft commander finally calls air traffic control, requesting a full stop landing at RAF Mildenhall. Air traffic control confirms the full stop landing will be on runway 29, and the crew perform their landing checks. The KC-135 landing speed is approximately 140 knots depending on its weight and flap settings, and Quid 78 lands and rolls to the very end runway and vacates before taxiing back to a parking spot. The aircraft commander will now communicate with the ground controller during its taxiing while the crew will carry out further landing and taxi checks.

Boeing KC-135R Stratotanker 58-0100 departs RAF Mildenhall in the summer sun.

Above left: Flying at their cruise altitude, the crew join ARA 4 and await the receivers.

Above right: The boomer takes his position and starts his checklist ready for the receivers.

Above left: The Boeing E-3 Sentry ZH104 looks to join for fuel with the refuelling flight.

Above right: The Boeing E-3 Sentry ZH104 is connected and taking in fuel from the KC-135R Stratotanker from the 100th ARW.

Above left: A General Dynamics F-16C from 480th Fighter Squadron joins and requests fuel from a Boeing KC-135R Stratotanker of the 100th ARW.

Above right: F-16C 96-0080 from the 480th Fighter Squadron from Spangdahlem Air Base, Germany, breaks away after receiving fuel from the KC-135R.

Above left: A 'Bone' flight F-15E Strike Eagle from the 494th Fighter Squadron heads towards the Boeing KC-135R Stratotanker requesting fuel.

Above right: A 'Bone' flight F-15E Strike Eagle from 494th Fighter Squadron joins up and begins his approach to the boom to take some fuel.

Above left: A 'Thud' flight F-15E Strike Eagle from 494th Fighter Squadron begin the joining process for fuel from the Boeing KC-135R Stratotanker.

Above right: F-15E Strike Eagle 91-0324 from 494th Fighter Squadron breaks away after taking some fuel from the KC-135R.

Above: A 'Rage' flight pair of F-15E Strike Eagles from the 494th Fighter Squadron, RAF Lakenheath, gracefully break away from the KC-135R.

Below left: Two F-15E Strike Eagles from 494th Fighter Squadron formate on the wing of the Boeing KC-135R Stratotankers before continuing with their flight.

Below right: Aircraft Commander Capt. J. Fuller Squadron crest.

A 'Quid 78' flight Boeing KC-135R Stratotanker on finals for runway 29 at RAF Mildenhall.

A view from the control centre of both RAF Mildenhall and RAF Lakenheath (RAPCON).

A view from the flight deck of the KC-135 Stratotanker.

A 'Quid 78' flight Boeing KC-135R Stratotanker taxies back to parking via taxiway Bravo at RAF Mildenhall.

The ground crew will be waiting at the parking area to marshal the tanker into stop position. As the aircraft comes to a stop and the chocks are put in place, the engines are shut down. Post-flight checks are completed before the aircraft is handed back to the squadron. The maintenance personnel will begin their checks on the aircraft to rectify any faults that the crew may have listed on their flight.

The crew would head back to debriefing room, where all aspects of their flight will be discussed and documented before other duties or going home. A life as a tanker pilot is unpredictable but exciting. It is full of long hours of flying and pushing papers when you are not in the air, but you get to see the world and serve your country while flying.

Above: Boeing KC-135T Stratotanker 59-1471 being marshalled in by ground personnel at RAF Mildenhall.

Next page
Above: Ground crew marshal the aircraft into the parking area, guiding the aircraft in using the marker batons.

Middle: With the aircraft safely stopped at the stop bar, ground personnel add the chocks to the wheels.

Below: A Boeing KC-135 Stratotanker parked peacefully on the dispersal until the next refuelling mission at RAF Mildenhall

U.S. AIR FORCE

U.S. AIR FORCE

6
Operational Wing: RAF Lakenheath

The 48th Fighter Wing (FW) also known as the 'Liberty Wing' is part of the USAF Third Air Force, and its headquarters are based in the UK at RAF Lakenheath in Suffolk. The 48th FW is home to the three squadrons of the F-15 Strike Eagles. Lakenheath is the only base in Europe to host the McDonnell Douglas F-15s and was given the name 'Statue of Liberty Wing' on 4 July 1954. It remains the only US Air Force unit with both a name and a numerical designation.

The 48th Fighter Wing supports operations for the USAF-AFAFRICA, US European Command, and NATO. The 'Liberty Wing' has over 5,500 active military personnel, plus 2,000 British and US civilian staff, which includes members at the sister station RAF Feltwell. The 48th also houses the 56th Rescue Squadron's HH-60G Pave Hawk Combat Search and Rescue helicopters. In 1952, the 137th Fighter-Bomber Wing at Chaumont Air Base, France, was redesignated the 48th Fighter-Bomber Wing. The fighter squadrons were renamed 492nd, 493rd, and 494th, and they are still around today.

F-84Gs and support aircraft were then assigned to the 48th Fighter-Bomber Group and the 48th FBW. They were set up to command all the roles for both the support and the fighter-bomber groups. The 48th FBW supported NATO and the USAFE. They were frequently deployed to Wheelus AFB, Libya, for training.

Through hard training, the 48th were developed into a premier fighter wing in France, and the staff of 48th worked hard to develop Chaumont Air Base, the result of which is rated as one of the best air bases in Europe. Its squadrons remained unchanged while flying three different type of fighters, the F-84G, F-86F, and the F-100D. With all this airpower, it gave the unit the capability to fight either a conventional or nuclear war, should the opportunity arise.

To improve French and American relations, the 48th Wing Staff had the idea to change the wing insignia. Chaumont Air Base is located not far from the workshops of Frédéric Bartholdi—the French architect who designed the Statue of Liberty. The contemporary design incorporated the Statue of Liberty, and throughout Europe, the 48th fittingly became known as the 'Statue of Liberty Wing'. On 4 July 1954, the mayor of the town of Chaumont bestowed the honorary title of the *Statue de la Liberté*.

In 1953, the wing was assigned the modern F-86F Sabre, then, in 1956, it had another upgrade to the F-100D 'Super Sabre'. The year 1957 saw the wing begin realigning its units as part of an Air Force worldwide reorganisation. By 1958, the 48th dropped the 'fighter-

bomber' title and became the 48th Tactical Fighter Wing. The three flying units also changed designation, becoming tactical fighter squadrons.

Unfortunately, international relationships between France and the US government deteriorated, resulting in French President Charles de Gaulle demanding the removal of NATO forces from the country in 1960. On 15 January that year, the 48th redeployed to an empty SAC heavy bomber base, better known as RAF Lakenheath, England. The wing's three fighter squadrons lifted off Chaumont's runway and, after making farewell passes over the outlying village, departed north towards the English Channel. A short while later, they all arrived at RAF Lakenheath for the first time. This was to be their home and base of operations.

In 1961, ever-increasing tensions with East Germany and the building of the Berlin Wall contributed to Cold War tensions hitting an all-time high. RAF Lakenheath served as a rotational base for the B-47 and B-52 aircraft throughout the Berlin Crisis and Cuban Missile Crisis. A piece of the Berlin Wall was dedicated on Lakenheath to remember the 'slain peacekeepers of the world'.

Between 1963 and 1972, the 48th Wing's F-100 fleet maintained its readiness. The fleet participated in many USAFE and NATO exercises, training to react to any aggression from the Soviet Union. The fleet underwent a series of NATO tactical evaluations, for which they earnt the wing its first Air Force Outstanding Unit Award. The wing conducted several deployments to Turkey, Italy, Spain, and across the UK.

In late 1971, the 48th TFW started its conversion to the McDonnell Douglas F-4D Phantoms II. The aircraft were transferred from the 81st TFW at RAF Bentwaters but the conversion to the F-4D took several years—the last F-100 finally departing in 1974. With the arrival of the Phantoms, the F-4s adopted a common tail code of 'LK'. This tail code lasted only a few months, and the 48th TFW was recoded to 'LN'. The F-4D carried squadron identifying fin cap colours of blue, yellow, and red—492nd, 493nd, and 494th—which are still present today.

Tactical squadrons of the 48th Operations Group are the 492nd Tactical Fighter Squadron (blue fin colours); the 493rd Tactical Fighter Squadron (yellow fin colours); and the 494th Tactical Fighter Squadron (red fin colours).

In early 1977, the F-4 transition to the newer Swing-Wing F-111s took place, and it was quick and without any significant problems. As a result, the 48th Wing was to receive its third Air Force Outstanding Unit Award for such a smooth transition. As quick as the transition

Two F-15E Strike Eagles from 49th Fighter Squadron formate on the wing of the Boeing KC-135R Stratotanker before continuing with their flight.

happened, the wing began a series of monthly exercises and deployments that took the Liberty Wing to Italy, Iran, Greece, and Pakistan.

By the mid-1980s, terrorists struck targets from the bombing of US Marines in Beirut to Berlin, from Rome to Rotterdam. Some of these attacks were influenced by the Libyan government, so, in retaliation, US President Ronald Regan ordered a strike against targets in Tripoli. Assigned for this mission, known as Operation El Dorado Canyon, were the F-111s of the 48th TFW. The F-111Fs from Lakenheath were chosen for their capability to fly long distances and deliver laser-guided munitions with fantastic accuracy. On 17 February 1987, Commander-in-Chief General Charles L. Donnelly Jr visited RAF Lakenheath presented decorations to those who participated in the operation.

Iraq's President Saddam Hussein directed his forces to invade Kuwait and threatened an invasion of Saudi Arabia. Within days of the invasion, US President George Bush directed the deployment of the US Armed Forces to Saudi Arabia in what became known as Operation Desert Shield. President Bush referred to Desert Shield as 'one of the most important deployments of allied military power since the Second World War'.

In response to the Iraqi invasion of Kuwait, Wing Commander Col. Thomas J. Lennon from 48th Tactical Fighter Wing received a phone call from the USAF headquarters. He was asked if the 48th Tactical Fighter Wing was ready to deploy, and on 25 August, eighteen F-111s took off from Lakenheath. This was significant as they were the first USAFE aircraft to deploy on Operation Desert Storm.

The 492nd, 493rd, and 494th flew to Taif Air Base, Saudi Arabia, supplying sixty-six F-111Fs. During the war, F-111Fs flew thousands of sorties, and their accuracy enabled them to successfully destroy Iraqi armour, artillery, bridges, military airfields, and command control posts.

On the 48th Tactical Fighter Wing's return to RAF Lakenheath in 1991, it had flown over 1,900 sorties and hit over 2,200 targets. By the mid-1990s, the USAF began restructuring and established several command positions, which was to include the 48th Logistics Group, 48th Medical Group, 48th Support Group, and 48th Operations Group. The program redesignated many of the USAF units by removing the 'tactical' from their names. So, in October 1991, the 48th Tactical Fighter Wing became the 48th Fighter Wing, the 492nd, 493rd, and the 494th becoming simply Fighter Squadrons. It was also decided in this period that the aircraft should be replaced again, changing from the F-111 to the F-15 Eagle. In 1992, the first F-15E landed at Lakenheath, with the last of the F-111F departing for units back the US in mid-December 1992.

The changes also saw the 493rd Fighter Squadron become deactivated in 1993, only to be reactivated a year later and receive the first maintenance trainer McDonnell Douglas F-15C/D Eagle. The 493rd received its twenty-second aircraft in July 1994, making them a full complement squadron. This marked the first time that the 48th had an air-to-air inventory after flying for more than fifty years.

The 48th Fighter Wing has been involved in a vast array of deployments since the early 2000s. With the terror attacks in the US on the Twin Towers, various units from the 48th Fighter Wing was deployed to Afghanistan and Iraq. The 48th FW supported key campaigns that included Operation Enduring Freedom, as well as Operations Iraqi Freedom and New Dawn. The 48th continued in active campaigns by participating in Operation Odyssey Dawn in Libya in 2011. Unfortunately, one of the Wing's F-15 aircraft crashed in Libya, but both aircrews survived with minor injuries.

The US Department of Defense announced in 2015 that the future of the 48th Fighter Wing and RAF Lakenheath would be prosperous for Lakenheath. The air base was chosen to become home to the first two squadrons of US Lockheed Martin F-35 Lightning II fifth-generation stealth fighter jets. The first of these new generation aircraft is due on base in 2020.

The 492nd Fighter Squadron

The 492nd Fighter Squadron was originally known as the 55th Bombardment Squadron back in 1941 and was redesignated to the 492nd Fighter-Bomber Squadron in 1943. It flew a vast array of aircraft, which included the B-18 and the P-47 during the Second World War. After deactivation in 1945, the squadron was reformed in 1952 at Chaumont Air Base, France, and received the F-86F Sabre and the F-100D, before moving to RAF Lakenheath in 1960.

The squadron was equipped with the F-4 Phantom and the F-111 during the 1970s and was finally given the F-15E in May 1992. Throughout its proud heritage and service, the 492nd has contributed to both NATO and USAFE assignments. In recognition of these contributions, the 492nd was named the Outstanding Fighter Squadron in USAFE for 1986.

In 1990, the squadron was deployed to Taif, Saudi Arabia, to support Operation Desert Shield. They were called into combat in 1991 at the onset of Operation Desert Storm, and during the campaign, the squadron flew over 800 missions, delivering more than 2.5 million pounds of munitions. In 1991, the 492nd were deployed to Incirlik Air Base, Turkey, in support of Operation Provide Comfort, and flew over 550 missions on behalf of the UN's inspection teams and the Kurdish relief effort. The squadron returned to RAF Lakenheath in 1992 and became the USAFE's first unit to transition to the F-15E Strike Eagle. The squadron was declared mission ready on 1 July 1993.

In 1993, the 492nd Fighter Squadron was again deployed to Incirlik Air Base, Turkey, in further support of Operation Provide Comfort, completing over 400 missions for the Kurdish relief effort. In 1994, the 492nd redeployed to Aviano Air Base, Italy, in support of Operation Deny Flight. In 1996, the 'Mad Hatters' played a vital role in stopping the Serb offensive in Kosovo and set the foundations for the NATO-led international peacekeeping force in Kosovo. The 492nd Fighter Squadron set the standard for 'anywhere, anytime'—this saying came about because the squadron was willing to pack it up and go wherever and whenever it was needed.

The squadron participated in Operation Odyssey Dawn in Libya in March 2011, along with numerous deployments to Southwest Asia supporting Air Expeditionary units as part of the ongoing Global War on Terrorism supporting Operation Iraqi Freedom and Operation Enduring Freedom.

Recent operations have seen the 'Mad Hatters' participate in a joint exercise conducted by NATO allies, Portugal, and the UK. Exercise Noble Arrow consisted of various air-to-air, air-to-surface, and air-to-sea scenarios that strengthened allied force communications. The 492nd Fighter Squadron recently was named the winner of the David C. Schilling Award. The award, sponsored by the Air Force Association, recognises 'the most outstanding contribution in the field of flight' in the atmosphere or space by an air force military member, air force civilian, unit, or group of individuals.

October 2017 saw the return of six F-15Es from the 492nd after being on deployment to the Middle East.

In March 2018, thirteen F-15E Strike Eagles and approximately 300 support personnel from assigned to the 492nd Fighter Squadron arrived at Andravida Air Base, Greece, to participate in exercise INIOHOS 18.

In May 2018, ten F-15E Strike Eagles from the 492nd participated in exercise Pointblank. Also in this month, the 492nd participated in the large force exercise Point Blank. Working alongside the RAF Typhoons based at RAF Lossiemouth, the 492nd gained vital training in both offensive and defensive operations.

In August 2018, the 492nd supported a RAF-led exercise, Typhoon Warrior, with Typhoon fighter jets based at RAF Coningsby. This was the RAF's premier air combat training exercise.

In October 2018, the squadron successfully completed a BAK-14 certification and close air support training with Estonian Joint Terminal Attack Controllers at Amari Air Base, Estonia.

On 30 January 2019, a F-15D assigned to the squadron was painted in the heritage colours of its Second World War P-47 Thunderbolt predecessor at RAF Lakenheath. The 48th Fighter Wing officially unveiled the aircraft publicly during a ceremony held at RAF Lakenheath on the 31st.

F-15E Strike Eagle 96-0205 from the 492nd Fighter Squadron, with full reheat, departs RAF Mildenhall after diverting due to an operational reason.

F-15E Strike Eagle 96-0202 from the 492nd Fighter Squadron taxies to active runway 24 at RAF Lakenheath on another training sortie.

F-15E Strike Eagle 98-0133 from the 492nd Fighter Squadron departs RAF Lakenheath, heading north to Scotland on a training sortie.

Above: Ground personnel work hard as a pair of F-15 Strike Eagles from the 492nd Fighter Squadron prepare to launch from RAF Lakenheath.

Below: The 'Mad Hatters'—the 492nd Fighter Squadron—parked on their dispersal at RAF Lakenheath.

F-15E Strike Eagle 97-0217 from the 492nd Fighter Squadron with ground personnel conducting final checks before departing RAF Lakenheath.

Above: F-15E Strike Eagle 91-0308 from the 492nd Fighter Squadron taking off with full afterburner from RAF Lakenheath.

Below: F-15E Strike Eagle 97-0220 from the 492nd Fighter Squadron departs RAF Lakenheath with full afterburner during a night sortie.

The 493rd Squadron

The 493rd Squadron was equipped with a variety of second-line aircraft back in 1942, with both single and twin engines, to prepare its pilots and maintenance crews for eventual combat. In 1944, the 493rd was assigned the P-47 Thunderbolt and was converted to an operational fighter squadron deployed to the European Theatre of Operations and assigned to the IX Fighter Command in England.

In 1944, the 493rd assisted with the Normandy invasion by dropping munitions on bridges and other high value targets, including gun positions, railway lines, and trains. In 1945, the squadron was sent to Laon in France, and after a few weeks in France, the squadron was ordered to return to the US. However, after the atomic bombs of Hiroshima and Nagasaki, and the sudden end of the Pacific War, the squadron was deactivated two months later as part of the massive post-war drawdown.

In 1952, the squadron was reactivated as a NATO fighter-bomber squadron based in France. In 1954, the squadron was equipped with the F-86F Sabre, conducting operational readiness exercises. In 1960, the squadron and the 48th Fighter Wing moved to RAF Lakenheath after disagreements with France over concerns about atomic storage and custody issues. At this point, the F-100D Super Sabre was assigned to the squadron.

In 1971, the squadron began to convert to the McDonnell Douglas F-4D Phantom II, with aircraft being transferred from the 81st Fighter Wing at RAF Bentwaters. In 1972, the 48th Fighter Wing changed its tail code from 'LK' to 'LN'. In 1977, the 493rd received General Dynamics F-111 Aardvarks and was sent out on deployments, which took the Liberty Wing to Italy, Iran, Greece, and Pakistan.

In 1986, the 493rd participated in Operation El Dorado Canyon, which was an air raid on Tripoli in Libya, and it was also involved in Operation Desert Storm in 1991. In 1994, the 493rd changed from the F-111 to the McDonnell Douglas F15C/D—this was the first time that the squadron had flown a specifically air-to-air aircraft; for over fifty years, it had flown an air-to-ground aircraft.

The squadron participated in Operation Odyssey Dawn in Libya in March 2011, along with numerous deployments to Southwest Asia supporting Air Expeditionary units as part of the ongoing Global War on Terrorism supporting Operation Iraqi Freedom and Operation

Enduring Freedom. In 2014, the 493rd was recognised as the USAF's top fighter squadron for its extensive work intercepting Russian aircraft in Baltic air space as tensions grew in the east.

In 2015, to support the United States' commitment to the fight against ISIL in Iraq and Syria, six F-15C Eagles from the 493rd was deployed to Incirlik Air Base, Turkey.

The Grim Reapers were part of the Icelandic Air Policing mission and they rapidly deployed to Lithuania to assist in the Baltic Air Policing missions against Russia as tensions grew between Russia and NATO allies. While on deployment, the 493rd flew over 2,800 hours and intercepted over thirty Russian aircraft. In April 2016, the USAF's F-22 Raptors deployed to RAF Lakenheath to conduct air training with other Europe-based aircraft. The aircraft belong to the 95th Fighter Squadron, 'The Boneheads', at Tyndall Air Force Base, Florida. They were met by F-15E Strike Eagles from Lakenheath and performed a touch and go at RAF Marham before heading to their destination, where they were deployed until May 2016.

In April 2017, F-35A Lightning IIs from the 34th Fighter Squadron at Hill Air Force Base, Utah, joined the F-15C Eagles from the 493rd Fighter Squadron while on deployment in Europe. The F-35A is a fifth-generation fighter that provides the joint war fighter unprecedented global precision attack capability against current and emerging threats.

In May 2017, twelve F-15C Eagles and approximately 200 personnel from the 493rd Fighter Squadron deployed to bases in Finland and Sweden for the Arctic Challenge 2017 exercise. The multinational training exercise aims at building relationships and increasing interoperability. The exercise included participants from the US, Finland, Sweden, Norway, the UK, Germany, France, Belgium, Switzerland, Canada, and the Netherlands, as well as representatives from NATO.

July 2017 saw the arrival of eighteen F-16 Fighting Falcons from the 52nd Fighter Wing, Spangdahlem Air Base, Germany, to participate in a flying training deployment. The 493rd and the 52nd Fighter Wing trained alongside the British during this pre-planned training engagement, which was scheduled to last several weeks.

August 2017 saw the 493rd deploy over 140 airmen and seven aircraft to Siauliai Air Base, Lithuania, to assume control of the NATO Baltic Air Policing peacetime collective defence mission.

October 2017 saw six F-22 Raptors belonging to the 27th Fighter Squadron and 94th Fighter Squadron, Joint Base Langley-Eustis, Virginia, join the 493rd for a three-week deployment, while another six arrived a week later to return stateside after being on deployment in the Middle East.

On 8 January 2018, the 493rd Expeditionary Fighter Squadron handed over the reins for NATO's Baltic Air Policing mission to Denmark at Siauliai Air Base, Lithuania.

In July 2018, Col. William Marshall assumed command of the 48th Fighter Wing from Col. Evan Pettus.

July 2018 also saw the USAF deploy approximately 280 Airmen and thirteen F-15C/D Eagles from the 493rd Expeditionary Fighter Squadron to Keflavik Air Base, Iceland. This was to conduct Air Surveillance missions in support of NATO alliance commitments.

In August 2018, the 493rd returned to RAF Lakenheath after over a month's deployment Reykjavík, Iceland. This was in support of NATO's Icelandic Air Surveillance mission.

In September 2018, airmen from the squadron completed exercise One Sky over the sovereign skies of Poland.

In October 2018, the 493rd participated in a series of large force exercises with F-22 Raptors assigned to the 1st Fighter Wing, Joint Base Langley-Eustis, VA, and F/A-18 Super

A pair of F-15C Strike Eagles from the 493rd Fighter Squadron hold on active runway 24 at RAF Lakenheath, awaiting air traffic control clearance.

F-15C from the 493rd Fighter Squadron breaks left for landing at RAF Lakenheath after completing a training sortie.

Above left: The Lockheed Martin F-22 Raptor from the 95th Fighter Squadron, Tyndall Air Force Base, graces the media during an event at RAF Lakenheath while on deployment.

Above right: Lockheed Martin F-22 Raptor 05-4106 from the 95th Fighter Squadron banks to the right, showing topside profile as it departs RAF Lakenheath on a training sortie.

Lockheed Martin F-22 Raptor 05-4089 from the 95th Fighter Squadron departs RAF Lakenheath on a training mission while on deployment.

Lockheed Martin F-22 Raptor 05-4086 taxies from the northern side of the airfield to active runway 24 at RAF Lakenheath for a training mission while on deployment.

Above: Lockheed Martin F-22 Raptor 05-4086 from the 95th Fighter Squadron on finals at RAF Lakenheath, returning after a training sortie with RAF Typhoons while on deployment.

Below: Air Commodore Ian Duguid OBE, Typhoon Force Commander RAF Coningsby, Major General Eric W. Vollmecke is Air National Guard Mobilisation Assistant to the Cdr Col. David Eaglin, Vice Commander of the 48th Fighter Wing at RAF Lakenheath. They stand proud with the F-15 Strike Eagle during the Lockheed Martin F-22 Raptor media day.

Lockheed Martin F-22 Raptor 04-4072 from the 95th Fighter Squadron, with full reheat, departs RAF Lakenheath, heading back to the States after deployment.

Above: Lockheed Martin F-35A 14-5094 from the 34th Fighter Squadron, Hill Air Force Base, on deployment at RAF Lakenheath with the 494th Fighter Squadron.

Below: Lockheed Martin F-35A 14-5098 from 34th Fighter Squadron, Hill Air Force Base, departs RAF Lakenheath while on deployment with the 494th Fighter Squadron.

The pilot and crew chief complete their final checks of the F-16C from the 480th Fighter Squadron during deployment at RAF Lakenheath.

Above: Seven F-16Cs from the 480th Fighter Squadron are parked up on the dispersal at RAF Lakenheath while on deployment in 2017.

Below: F-16C 90-0827 from the 480th Fighter Squadron departs RAF Lakenheath on a training mission with the 493rd Fighter Squadron.

Above left: Four F-15Cs from the 493rd Fighter Squadron on the 'Last Chance', awaiting departure clearance.

Above right: An F-15C from the 493rd Fighter Squadron taxies to the active runway before heading out on a night training sortie.

Above: F-15C 86-0164 from the 493rd Fighter Squadron takes off during a night mission.

Below: Lockheed Martin F-22A Raptor 09-4185 from the 1st Fighter Wing departs RAF Lakenheath on the final leg home back to the States.

Lockheed Martin F-22A Raptor 09-4175 from the 1st Fighter Wing joins five other F-22A Raptors on their journey back to the States.

Above: Six F-15C Strike Eagles from the 493rd Fighter Squadron hold at the 'Last Chance' before a mass launch at RAF Lakenheath.

Below: A formation of three F-15Cs from the 493rd Fighter Squadron, RAF Lakenheath, flypast the Duxford Air Museum in a show of strength.

Hornets assigned to Carrier Air Wing One, deployed from the *Nimitz*-class aircraft carrier USS *Harry S. Truman*.

In November 2018, the 493rd Fighter Squadron participated in the NATO Tactical Leadership Programme 18-4 at Amendola Air Base, Italy.

The 494th Fighter Squadron

The 494th Fighter Squadron (better known as the 'Panthers') continually trains as a combat ready fighter squadron, capable of executing strategic attacks and counterattack missions in support of USAFE and NATO operations.

The squadron has a vast array of munitions used for air superiority and surface attacks, plus the most advanced precision guided weapon. The squadron can be deployed to any theatre of operations in the world, at any time.

The 'Panthers' consider themselves the 'finest, baddest, and meanest' fighter squadron in the air force, with their motto being 'making our enemies bleed red'.

The 494th Fighter Squadron has an extensive history that stems from the days of the Second World War, with the unit flying a variety of aircraft, including the P-40 Warhawk, A-35 Vengeance, and the F-86 Sabre, the F-4 Phantom, the F-111 Aardvark, and finally the F-15E Strike Eagle.

In 1944, the 494th was deployed to Europe and the squadron was awarded nine battle honours in recognition of combat service and sacrifice during the war. The squadron was based at RAF Ibsley and flew escort and fighter cover missions to Europe to prepare regions for the ground assaults. During the Normandy invasion, the 'Panthers' provided heavy support to Allied forces and engaged in direct combat against German forces. The 494th had a very good reputation for effective close air support with its P-47s; it also conducted missions against factories, bridges, and other depots within Germany, which helped slow its war machine projects.

During the Battle of the Bulge in 1944, the 'Panthers' were involved in its heaviest combat of the war with the campaign lasting over six weeks. The 494th provided close air support operating just above the trees, helping the aircraft avoid detection; the unit also destroyed multiple enemy vehicles and buildings. In total, the 48th Fighter Group flew over 1,400 sorties and exhausted the German resources during the Battle of the Bulge. Subsequently, in March 1945, the 494th Fighter

Squadron also played a key role in Operation Plunder Varsity, the combined land and airborne assault across the River Rhine. Flying from an occupied airfield at Kelz, Germany, pilots of the 48th Fighter Group launched 189 sorties, fired up to 70,000 rounds of munitions, and dropped more than 500 bombs. In recognition for its combat operations supporting the advancing US Army across the heart of the continent, the 'Central Europe' streamer was awarded to the 494th.

After the Second World War, the 494th headed back to the US and Seymour-Johnson Field, North Carolina, but, with tensions running extremely high with the Soviet Union in the '50s, the 'Panthers' received orders to return to Europe. In 1952, the squadron was assigned to Chaumont Air Base, France, flying the F-84 Thunderstreak, the F-86 Sabre, and the F-100 Super Sabre.

In 1960, the 494th was on the move again, this time to the UK, based at RAF Lakenheath—the home airbase to the Panthers to the current day. During the early 1970s, no aircraft or personnel were assigned to the 494th, but it returned to operational status in 1974 when it was assigned the McDonnell Douglas F-4 Phantom II, which was in use until 1977, when the wing re-equipped to the General Dynamics' F-111.

In 1986, the 494th was called upon again due to the foreign threats in the world and the F-111s from 48th Fighter Wing attacked targets in Tripoli, Libya, aimed to eliminate the ability of the country to support and train international terrorists. In 1990, the invasion of Kuwait by Iraq saw the largest mobilisations since the Second World War by the US, codenamed Operation Desert Shield and Operation Desert Storm.

The US, the UK, and a large coalition of other nations demanded the total withdrawal of Iraqi forces from Kuwaiti territory, and in August 1990, the 48th Tactical Fighter Wing deployed to Taif in Saudi Arabia. Iraq failed to comply with the UN demands to leave Kuwait, and with that, Iraq felt the full brunt of the coalition's military offensive, better known 'Operation Desert Storm'.

The F-111s participated in many missions, unleashing their precision guided missiles and destroying bridges, airfields, and tanks. Five F-111s flew their only day mission of the campaign and, using GBU-24 bombs, struck five buildings in Kuwait City, which housed the Iraqi high command, and this action severely disrupted Iraq's command and control over its military forces within Kuwait.

In total, the Liberty wing flew nearly 2,000 combat sorties during 'Operation Desert Storm' and destroyed over 2,000 targets. In recognition of their contribution towards victory over Iraq during Operation Desert Storm, 'Liberation and Defence of Kuwait' streamer was awarded to the 494th Fighter Squadron.

In 1991, the 494th unit received its current designation of 494th Fighter Squadron, and in 1992, it was given a new aircraft, the McDonnell Douglas F-15E Strike Eagle, plus a restructure that reassigned the 494th from the wing to the 48th Operations Group, where it remains today.

In 1993, the 494th Fighter Squadron deployed to Incririk Air Base in Turkey and its mission was to provide counter air and reconnaissance missions. The F-15 Eagles, with their air-to-air capability, enforced a no-fly zone and protected high-value assets over northern Iraq.

In 1995, the 494th led NATO's first and largest airstrike against Bosnian Serb Command and Control facilities. In 1999, the 'Panthers' deployed to Aviano Air Base, Italy, where they were to participate in 'Operation Allied Force'. Allied Force was designed to stop the Serbian repression on Kosovo, and during this conflict, the 494th led at least half of the NATO strike missions. In recognition of service during the Noble Anvil campaign, the 'Air Force Outstanding Unit Award' streamer was conferred upon the 494th Fighter Squadron.

F-15E Strike Eagle 98-0132 from the 494th Fighter Squadron departs RAF Lakenheath with full reheat on a night training sortie.

Above: F-15E Strike Eagle 98-0132 from the 494th Fighter Squadron taxies to the active runway at RAF Lakenheath for night training sortie.

Below: F-15E Strike Eagle 91-0310 from the 494th Fighter Squadron takes off from RAF Lakenheath on an early morning mission.

In 2006, the 'Panthers' were tested again as they departed RAF Lakenheath heading for the Middle East, in support of 'Operation Iraqi Freedom' and Operation 'Enduring Freedom', with the F-15 Strike Eagles flying combat sorties over Iraq. The F-15s were assigned to attack key military command and control targets and they subsequently helped in close air support missions.

The 494th returned to RAF Lakenheath in 2007 after a very long deployment in Iraq. On Christmas Eve 2009, the eighteen F-15 Eagles from the 494th began its deployment to Bagram Air Base in Afghanistan. The F-15 Eagle was known as 'Thunder' by the Taliban. During this deployment, the 'Panthers' flew nearly 2,000 combat sorties and over 7,500 combat hours. The deployment was a remarkable success and proved why the 494th will always be known as the finest, baddest, and meanest.

The 494th Fighter Squadron was deployed in 2014 to the US Central Command (USCENTCOM), which has an area of responsibility that includes countries in the Middle East, North Africa, and Central Asia—most notably Afghanistan and Iraq. While the 494th were deployed, the unit flew over 1,500 missions.

In September 2015, the 'Panthers' participated in the month-long multi-lateral training for NATO Tactical Leadership Programme at Los Llanos Air Base, Spain.

In September 2016, the 494th Fighter Squadron deployed overseas in support of Tactical Leadership Programme 16-3 at Los Llanos Air Base, Spain. The training prepares NATO and allied forces' flight leaders to be mission commanders, lead coalition force air strike packages, instruct allied flying and non-flying personnel in matters related to tactical composite air operations, and provide tactical air expertise to NATO agencies.

In May 2017, the unit deployed in support of exercise Juniper Falcon to Uvda Air Base, Israel. Juniper Falcon 17 represents the combination of several bi-lateral component and Israeli Defence Force exercises that have been executed annually since 2011. These exercises were combined to increase joint training opportunities and capitalise on transportation and cost efficiencies gained by aggregating forces.

F-15E Strike Eagle 01-2000 from the 494th Fighter Squadron taxies to the northern side of the airfield to wait for his colleagues.

No. 97-0216 from 492nd Fighter Squadron departing RAF Lakenheath on a cold and snowy morning mission.

Eight F-15E Strike Eagles from the 494th Fighter Squadron parked on the frontline at RAF Lakenheath.

91-0326 from the 494th Fighter Squadron, with full reheat, takes off from RAF Lakenheath during an early evening training mission.

Above: F-15E Strike Eagle 91-0309 from the 494th Fighter Squadron departs RAF Lakenheath on another training sortie, heading to Scotland.

Below: Three F-15E Strike Eagles from the 494th Fighter Squadron sit at the 'Last Chance', awaiting further air traffic control instructions at RAF Lakenheath.

F-15E Strike Eagle 91-0326 from the 494th Fighter Squadron applies full reheat as it departs RAF Lakenheath on a late afternoon sortie.

Above: F-15E Strike Eagle 01-2001 from the 494th Fighter Squadron lifts of another navigational mission from RAF Lakenheath.

Below: F-15E Strike Eagle 01-2001 from the 494th Fighter Squadron taxies to the 'Last Chance' at runway 24, RAF Lakenheath.

In October 2018, airmen and supporting units of the 494th Fighter Squadron were reunited with family and friends after returning from a six-month deployment to an undisclosed location in Southwest Asia.

The 56th Rescue Squadron

The 56th Rescue Squadron (RQS) are known as the 'Jolly's'. It operates HH-60G Pave Hawk helicopters, conducting search and rescue missions, and it has recently been announced that the squadron will relocate to Aviano Air Base, Italy, starting in 2017.

The 56th Rescue Squadron is a combat-ready search and rescue squadron that can execute all-weather search and rescue missions, day or night, in many varied hostile environments in support of USAFE, USEUCOM, and NATO operations. The unit is ready to deploy at any time to anywhere in the world.

In 1952, with H-5 Dragonfly helicopters and SA-16 Albatross amphibious flying boats at Sidi Slimane Air Base, the 56th Air Rescue Squadron was activated and was a part of the French Moroccan 7 Air Rescue Service, providing search and rescue and medical evacuation services in North Africa and Southern Europe.

In 1953, the squadron converted to SH-19 Chickasaw helicopters and added SC-47 Skytrains, modified for ASR, to its inventory. The squadron was reassigned to the 12th Air Rescue Group in 1956 and then to the Air Rescue Service in 1958; attached to Detachment 3, 8th Air Rescue Group European Rescue Operations.

In 1971, three Sikorsky HH-3E Jolly Green Giants arrived on NAS Keflavik to become a permanent fixture. The unit's first mission was an Icelandic Lifesaving Association request for the airlift of a nine-year-old boy suffering from epileptic seizures. The squadron carried out combat search and rescues for airborne mission control and aerial refuelling in support of operations over Vietnam until 15 August 1973. It also performed local search and rescue missions until deactivated in October 1975.

The 56th Rescue Squadron was reactivated in 1988 on NAS Keflavik, Iceland, assigned to the 39th Special Operations Wing, Eglin AFB, Florida, until late 2006. It was also heavily involved in overseas operations that saw the 56th HH-60G Pave Hawk aboard USS *Ponce* during Operation Odyssey Dawn off the Libyan coast in 2011.

A pair of Sikorsky HH-60G Pave Hawks from the 56th Rescue Squadron arrive back at RAF Lakenheath after a training sortie.

Above: A Sikorsky HH-60G Pave Hawk 89-26205 hover taxies to the active runway at RAF Lossiemouth during Exercise Joint Warrior.

Below: HH-60G Pave Hawk 91-26353 taxies out to the threshold at RAF Lossiemouth during Exercise Joint Warrior.

The air, maintenance, and operations crews are alert twenty-four hours a day, 365 days a year, with the crew constantly training. In 1992, the 56th RQS was the only unit to convert from the HH-3 Jolly Green Giant to the HH-60G Pave Hawk and maintain a continuous SAR alert status.

The 56th RQS is a three-time recipient of the Jolly Green Association Mission of the Year Award. Due to the illustrious history of saving lives in and around Iceland, the 56th RQS was the first non-Icelandic organisation to be recognised for its outstanding rescue efforts by the Icelandic Seaman's Day Committee on 11 June 1995.

The 56th Rescue Squadron relocated to RAF Lakenheath and was reassigned to 48th Fighter Wing in June 2006 to provide Combat Search and Rescue CSAR missions in all weathers, day or night, deploying to hostile environments in support of the USAFE and NATO.

The squadron deployed to Bagram Air Base, Afghanistan, in 2010, conducting combat search and rescue missions. Rescue units in Iceland have aided citizens of over twenty countries, including the US, Soviet Union, the UK, Denmark, Sweden, Norway, Germany, and Iceland.

The 56th Rescue Squadron can be seen regularly at RAF Lakenheath conducting training missions. The unit uses the training grounds around STANTA and the NOOK. Unfortunately, one of the Pave Hawks was involved in a terrible incident in 2013 and all souls on board perished.

Recent deployments saw the 56th RQS participate in exercise Salamander, France, in November 2016, working alongside the French Air Force.

RAF Lakenheath, like her sister base RAF Mildenhall, is a wing known as the Liberty Wing, which has definite missions with specific goals. The wing would be made up of groups consisting of several squadrons and is usually commanded by a colonel; the base commander who has recently returned to the Liberty Wing is Col. William L. Marshall. Prior to assuming his current position, Col. Marshall was the Commander of the 332nd Expeditionary Operations Group. In this position, he was responsible for an F-15E Combat Fighter Squadron, a MQ-9 Squadron, a KC-135 Squadron, a Space Control Squadron, and an Operational Support Squadron conducting operations at an undisclosed location in Southwest Asia. Col. Marshall has flown in support of Operation Enduring Freedom, Operation Southern Watch, Operation Iraqi Freedom, and Operation Inherent Resolve. He has over 650 hours of combat in the F-15E.

RAF Lakenheath is the premier McDonnell Douglas F-15 Eagle fighter jet air base in Europe and the Liberty Wing take pride on being the best. Like the nearby sister base, RAF Lakenheath hosts other US fighter squadrons and NATO allied nation squadrons. These squadrons are either transiting through the UK, heading to the US, or coming from the States and heading further south into Europe or to the Middle East on deployment.

The 48th Fighter Wing's vision is to be its nation's premier combat wing, capable of dominating any adversary. Its objectives to sharpen combat capabilities, strengthen the team, and shape the installation requires a unique form of teamwork; therefore, personnel are proudly known as simply the 'Liberty Wing'.

The 48th Medical Group

The 48th Medical Group (MDG) provides operational medical support to the US European Command, US Africa Command, and US Central Command. The 48th MDG supports the 48th Fighter Wing at Lakenheath, the 100th Air Refuelling Wing at RAF Mildenhall, the 501st Combat Support Wing at RAF Alconbury, and the 352nd Special Operations Wing also based

HH-60G Pave Hawk 91-26353 taxies back to the dispersal after another successful mission during Exercise Joint Warrior.

Above: US Marine Corps' F/A-18 Hornet 163166 from the squadron VMFA-314, known as the 'Black Knights', based at Marine Corps Air Station Miramar, California, departing RAF Lakenheath to head home to the States.

Below: US Marine Corps' F/A-18 Hornet 162442 from the squadron VMFA-314, known as the 'Black Knights', based at Marine Corps Air Station Miramar, California, awaits to depart RAF Lakenheath to head home to the States.

USAF General Dynamics F-16C 91-0403 from 480th Fighter Squadron, Spangdahlem, departs, heading back to Germany after spending time at RAF Lakenheath while on deployment.

Fairchild Republic A-10 Thunderbolt II 78-0651 from the 354th Fighter Squadron, Davis-Monthan Air Force Base, departs home from RAF Lakenheath early morning in March 2015.

Seymour Johnson F-15E 89-0488 from the 336th Fighter Squadron, 'The Rocketeers', arriving at RAF Lakenheath on its way to the Middle East.

Above: Seymour Johnson F-15E 89-0501 from the 336th Fighter Squadron, 'The Rocketeers', arriving at RAF Lakenheath on its way to the Middle East

Below: Seymour Johnson F-15E 89-0492 from the 336th Fighter Squadron arriving at RAF Lakenheath on its way to the Middle East on deployment

at RAF Mildenhall. The 48th MDG primary and specialty health service provides care for more than 35,000 eligible US personnel and their spouses, totalling over 149,000 outpatients and 2,000 inpatient consultations annually at six operating locations throughout the UK and Norway. The 48th MDG has also completes 67,000 dental operations, over 184,000 laboratory procedures, 2,000 surgeries, 1,600 physiotherapy consultations, and over 450 new-born deliveries per year. The 48th MDG are simply known as 'Liberty Medics, A Cut Above!'

The 48th Maintenance Group

The 48th Maintenance Group, which includes six squadrons, is responsible for maintaining McDonnell Douglas F-15s, including engines, munitions, and support equipment for all three fighter squadrons and the helicopter squadron based at RAF Lakenheath.

The 48th Aircraft Maintenance Squadron

The 48th Aircraft Maintenance Squadron can be deployed worldwide at any time and administers the inspections and maintenance of the F-15E aircraft. The 48th Aircraft Maintenance Squadron (AMS) sustains two fighter squadrons in support USAFE, EUCOM, and NATO commitments.

The 748th Aircraft Maintenance Squadron

The 748th Aircraft Maintenance Squadron can be deployed worldwide and is one of the most distinct squadrons in the USAF. It includes two unique units, the 493rd Aircraft Maintenance Unit (AMU) and the 56th Helicopter Maintenance Unit. Each maintains a fleet of F-15C fighter aircraft and HH-60G Pave Hawk helicopters. The 493rd AMU maintains the fleet of McDonnell Douglas F-15Cs to maintain and accomplish air superiority and air defence missions. The 56th AMU is the only USAFE-based rescue squadron and maintains its fleet in support of disaster relief emergency operations.

The 48th Component Maintenance Squadron

The 48th Component Maintenance Squadron conducts maintenance and accessory repair of all F-15C, F-15D, F-15E, and HH-60G aircraft. The squadron also sustains and supports US Africa Command, CENTCOM, EUCOM, NATO, and USAFE tasking by performing equipment calibrations and intermediate level maintenance on F-15C, F-15D, F-15E, and C-130 avionics line replaceable units.

The 48th Equipment Maintenance Squadron

The 48th Equipment Maintenance Squadron supports the 48th Fighter Wing, USAFE, EUCOM, and NATO commitments by providing maintenance for one F-15C/ F-15D fighter squadron and two F-15E fighter squadrons. It is accountable for the maintenance of aerospace ground equipment, including runway lighting, structural repair to all building on the base, fabrication, corrosion prevention, aircraft inspection, and crash recovery.

The 48th Maintenance Operations Squadron

The 48th Maintenance Operations Squadron maintains the 48th Wing combat-ready status and aircraft fleet management by control, scheduling, and training of the wing's F-15C, F-15D, and F-15E aircraft and personnel.

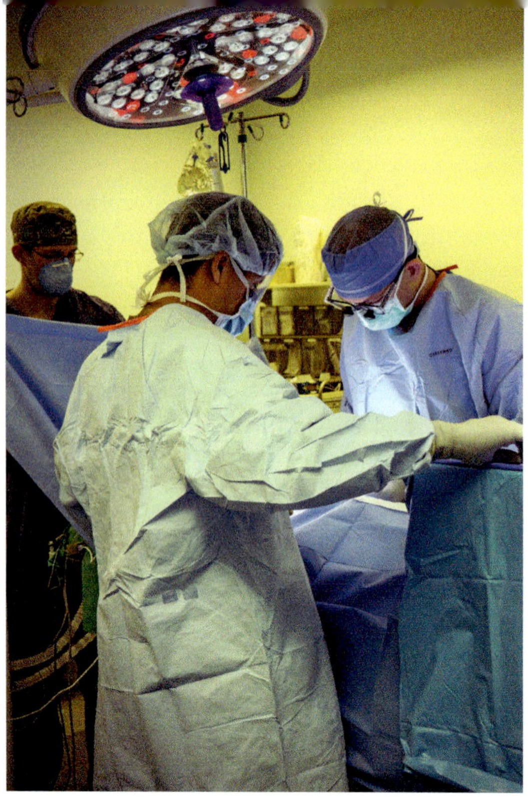

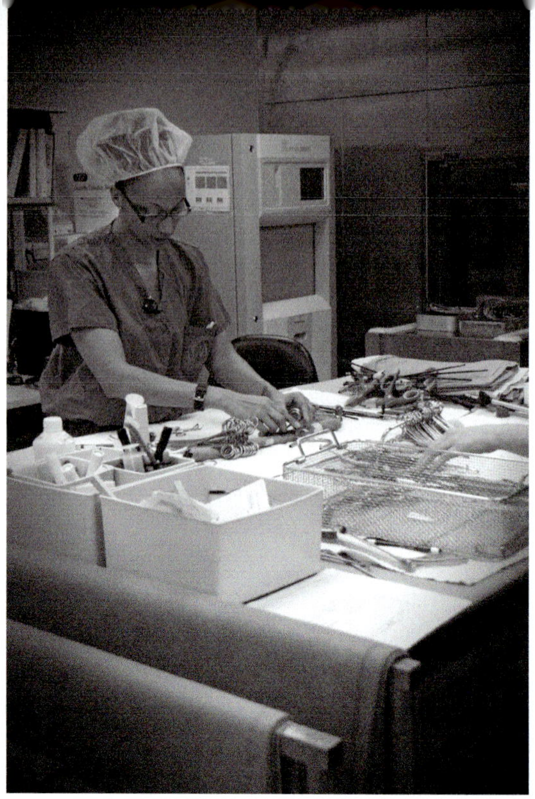

Above left: The 48th Medical Group (MDG) surgical team operate on personnel at the on base hospital, RAF Lakenheath.

Above right: The 48th MDG medical staff clean and replenish hospital and operating utensils.

Below left: The 48th MDG laboratory technician analyses biological specimens of patients.

Below right: The 48th MDG medical nurses administer care and attention to long term sick patients.

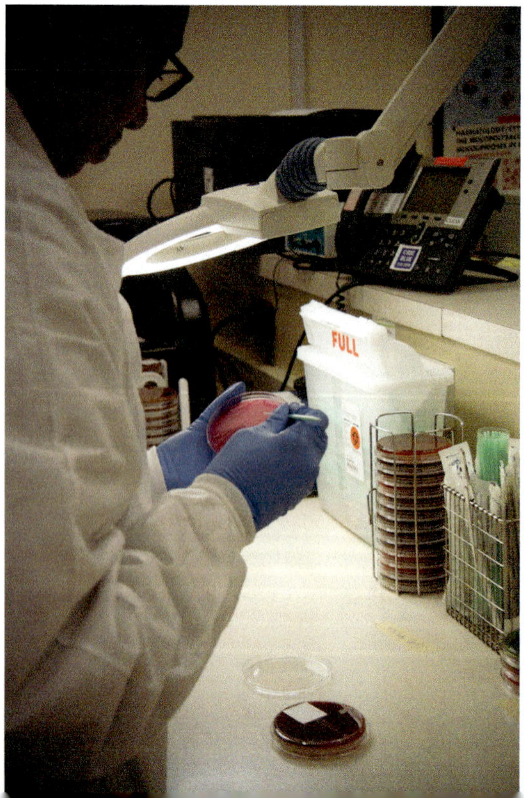

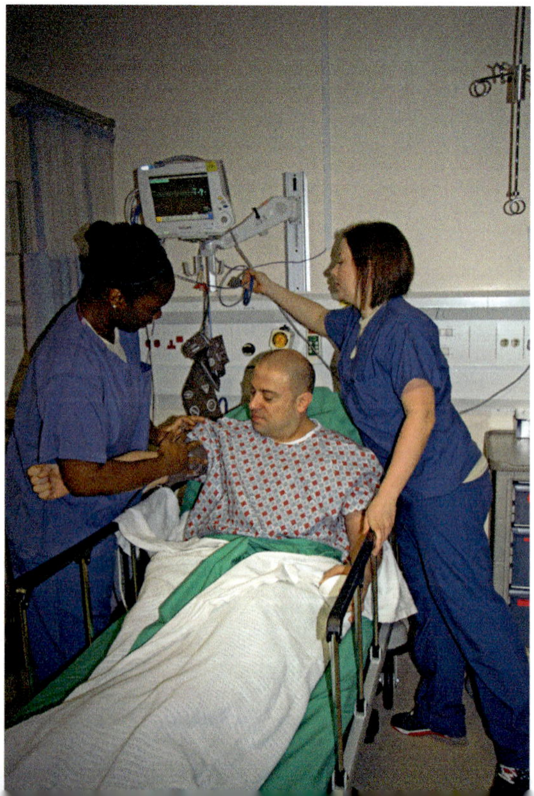

Above: Impressive view of the F-15 workshop at RAF Lakenheath.

Below left: The Pratt & Whitney F100 powers the F-15s.

Below right: The HUD is basically a piece of glass located on top of the pilot's dashboard. Calibration testing is vital to keep both aircraft and pilot safe in the skies.

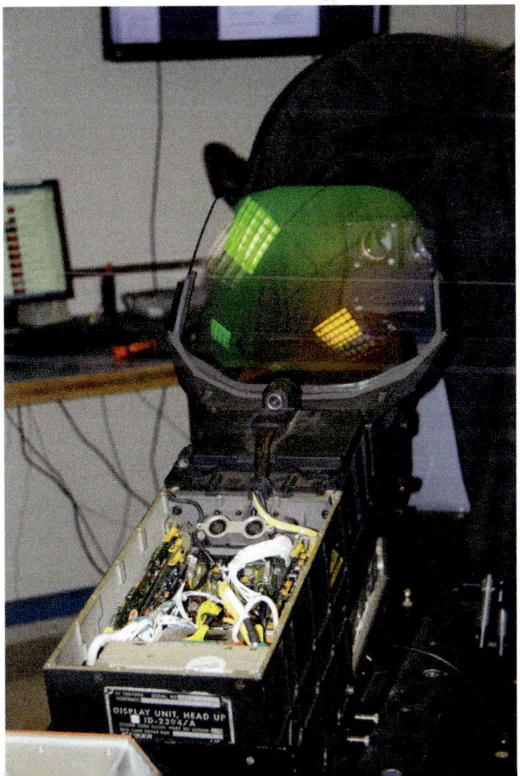

The 48th Munitions Squadron

The 48th Munitions Squadron provide all munitions and armament systems, including bombs, 20-mm rounds, countermeasures, marine markers, and small arms for all the F-15C, F-15D, and F-15E fighter squadrons supporting the 48th Fighter Wing, USAFE, EUCOM, and NATO commitments. It provides maintenance of critical equipment and weapons storage and security systems.

The 48th Operations Group

The 48th Operations Group (OG) consists of six squadrons, including those using the F-15C/D/E Eagle and Strike Eagle fighter aircraft and HH-60G Pave Hawk helicopters. The 48th OG also provides support personnel capable of accomplishing fighter and rescue operations worldwide and maintaining the combat-ready status for US European Command, US Africa Command, and NATO operations. The 48th OG delivers training, equipment, scheduling, weather intelligence, and command and control for all flying operations.

The 48th Operation Support Squadron

The 48th Operations Support Squadron (OSS) provides the operational structure for five of the F-15 squadrons plus the HH-60G Pave Hawk helicopters. The 48th OSS can support operations for USEUCOM, USAFRICOM, and NATO. The 48th OSS sustains the five operational flying squadrons by support, which includes flight and airfield management, air traffic control, weather intelligence, weapons and tactics, intelligence, scheduling, training, and group administration.

The 48th Mission Support Group

The 48th Mission Support Group (MSG) administers the 48th FW personnel, logistics, communications, contracting, engineering, and services support squadrons. The 48th MSG delivers fire protection for on base infrastructure, facilities, and off-base protection for personnel and home locations. The 48th MSG enrich the quality of life and service to over 15,000 military, civilian, and dependent personnel assigned to RAF Lakenheath, RAF Feltwell, and the surrounding areas.

The 48th Civil Engineer Squadron

The 48th Civil Engineer Squadron (CES) supports the largest European US Fighter Wing with a 530-personnel-strong squadron, providing guidance and aid to over 15,000 military, civilian, and family members at two locations. The 48th CES also provides emergency management, including aircraft emergencies—this could be on the ground or an aircraft declaring an inflight emergency (IFE). Once an IFE has been declared by the pilot, all the relevant emergency departments (including fire and arrester hook systems) are informed and prepare for the arrival of the aircraft. The 48th CES is heavily involved in both sections and is well trained for any emergency. The arrester cable (also known as 'catching the wire') is a permanent system that is installed on nearly all US airfields that operate fighter jets. The arresting cables or wires are flexible steel cables that span across the runway roughly 2 inches above and with rubber doughnuts to cushion the cable with the runway. The cable is engaged by the arresting hook of an incoming aircraft. The system is a hydraulic system that gradually brings the aircraft to a stop. RAF Lakenheath handle twelve to fifteen cable engagements per year.

Above left: Ground crew attack munition hooks to the F-15 before arming the aircraft.

Above right: Explosives without the warhead wait to be armed in the munitions section at RAF Lakenheath.

Above left: Paveway missiles loaded and secure on a transport trailer, before being transported to an awaiting F-15.

Above right: One of many warehouses that houses vital goods, whether it is household goods or engine parts.

Below: Fire tenders attend a practice fire drill to an aircraft fire in a Permanent Aircraft Shelter (PAS).

Above left: The fire crew, armed with water hose and breathing apparatus, is ready to tackle the aircraft fire.

Above right: Smoke billows out of the HAS. The lead firefighter gives the signal that he is ready to tackle the aircraft fire, overseen by the fire chief.

Above: Fire crew tackle the aircraft fire inside the HAS, with smoke billowing out.

Below: With the fire extinguished, the fire crew access the damaged caused by the fire.

RAF Lakenheath's arresting cables or wires span across the runway roughly 2 inches above and with rubber doughnuts to cushion the cable on the runway.

Cable huts are located at each of the runway, housing the cable generators.

The 48th CES provides environmental protection on air quality, land quality, waste and noise, and their effects. Airfield management can include runway lighting, snow clearing, and emergency and electrical faults, plus ordnance disposal with the process by which hazardous explosive devices are rendered safe. The 48th CES acquires, maintains, and operates over 600 official buildings and over 1,300 housing units, an on-base hospital, and five Department of Defense Dependent Schools covering over 2,500 acres.

The 48th Force Support Squadron

The 48th Force Support Squadron (FSS), a 700-personnel-strong squadron, is responsible for the implementation of all military personnel educations and professional development, airman and family readiness, career assistance, and first term airman centre services. It administers the airman leadership school for the 48th FW, the 100th ARW, and 352nd SOG, plus six separate units in various locations in the UK. The 48th FSS strengthens airman combat readiness, retention, and morale through programs promoting welfare and quality of life. The FSS oversee dining, lodging, fitness, honour guard, leisure activities, family programs, and funeral arrangements.

The 48th Security Forces Squadron

The 48th Security Forces Squadron (SFS) protect and defend the 48th FW both on base and while deployed. The 48th SFS is responsible for the protection of over 15,000 military and civilian personnel at two locations, plus coordinates law enforcement operations from high-ranking dignitaries to observing the local community of plane spotters along the perimeter fence with host nation's police departments. The 48th SFS's responsibilities include police patrols, gate guarding, flight line security, alarm monitoring, combat arms training and maintenance, and K-9 dog handling and patrols.

The 48th SFS operates the UK's only USAF correctional facility, which can hold up to sixteen inmates at any one time. The correctional facility manages, supervises, and directs inmates and provides for the health and welfare of detainees. The correction facility administers and coordinates the transfer of inmates with overseeing the operation of the facility. Inmates will conduct themselves with proper military behaviour while assigned to the correction facility. The 48th SFS occupies the largest combat arms training range, providing small arms weapons training to three USAF wings: the 48th FW, the 100th ARW from RAF Mildenhall, and the 501st CSW from RAF Alconbury.

The 48th Communications Squadron

The 48th Communications Squadron (CS) provides the 48th FW command, control, and communication systems, supporting both on base and while deployed. The 48th CS maintains air traffic control, meteorological, navigation, and information management, plus force protection and personal communications systems across the base. The 48th CS utilise over $64 million in communication assets, providing superior computer, data, and voice capabilities to USAFE's largest fighter wing and multiple command locations globally.

The 48th Contracting Squadron acquires goods and services for the Department of Defense in the UK, including the 48th FW, the 100th ARW, and 352nd SOW, and the Department of Defense Dependent's Schools furthermore provides the second largest Government Purchase Card (GPC) program in the USAFE, supporting over 25,000 personnel. It also deploys personnel worldwide for combat and humanitarian operations.

Armed Security Forces personnel guarding the aircraft from the 48th Fighter Wing, RAF Lakenheath.

Security Forces truck patrols the 494th Fighter Squadron dispersal at RAF Lakenheath.

Above left: Armed Security Forces personnel guard the main gates into and out of RAF Lakenheath.

Above right: All personnel wishing to gain entrance must have photographic ID. The Security Forces check everyone's ID before allowing entrance to RAF Lakenheath.

Above left: Security Forces demonstrate the command and strength of the K-9 section.

Above right: The K-9 handler of the security forces demonstrate why the K-9 section is vital for the security of RAF Lakenheath.

The K-9 handler and his dog stand proud outside the detention centre at RAF Lakenheath.

Sergeant Denis stands proud with his colleagues outside the RAF Lakenheath correctional facility.

The view from a busy air traffic control at RAF Lakenheath.

Radar Approach Control System (RAPCON) at RAF Lakenheath, keeping the skies over Suffolk safe.

The 48th Logistics Readiness Squadron

The 48th Logistics Readiness Squadron (LRS) (also known as the 'Saxons') is split into numerous sections called flights. These all have a vital role that help the 48th LRS provide a full range of logistical competence for seventy-six MacDonnell Douglas F-15C/E aircraft and five HH-60G Pave Hawk helicopters assigned to the 48th FW, which supports US European Command, US Africa Command, and NATO commitments. The 48th LRS stores, inspects, and distributes supply assets worth more than $900 million dollars with $22 million worth of petroleum and $48 million on vehicles. The 48th LRS provides vehicle operations functions and is responsible for providing transportation services to support the mission, dispatch, pick-up, and delivery operations, and vehicle and equipment support. The 48th LRS also maintains and manages all the installation's motor vehicle fleet.

The 48th LRS has one of the most vital missions in the 48th FW: Fuels Management Flight. This is to ensure the jet fuel is ready for fuelling all aircraft, government-owned vehicles, and other gas-operated equipment.

The Fuels Management Flight conducts, on average, fourteen hot pit refuelling operations per month, issuing an average of 280,000 gallons of fuel—hot pit refuelling reduces aircraft ground time and cuts aircraft regeneration time by 50 per cent.

The High Mobility Multipurpose Wheeled Vehicle (HMMWV), commonly known as the Humvee, undertaking routine maintenance checks.

The LRS de-icing truck stowed away until the winter months, when it will be used daily.

F-15E from the 494th Fighter Squadron being refuelled from within the PAS at RAF Lakenheath.

F-15E from the 494th Fighter Squadron being refuelled outside the PAS at RAF Lakenheath.

Above: F-15E from the 494th fighter Squadron simulated hot pitting outside the hardened aircraft shelter (HAS) at RAF Lakenheath.

Below: Ground personnel simulate hot pit refuelling of the F-15E from the 494th Fighter Squadron outside the HAS at RAF Lakenheath.

7
RAF Lakenheath-Based Aircraft

F-15 Strike Eagle

The McDonnell Douglas (now Boeing) F-15E Strike Eagle is an American all-weather multirole strike fighter, designed by McDonnell Douglas to gain and maintain air supremacy in aerial combat. It is among the most successful modern fighters, with over 100 victories and no losses in aerial combat.

The USAF selected McDonnell Douglas's design in 1967 to meet the service's need for a dedicated air superiority fighter and it entered service in 1976. The USAF have a vast array of squadrons of F-15s, but the F-15E Strike Eagles can be distinguished from other USAF Eagle variants by darker aircraft camouflage and conformal fuel tanks mounted along the engine intake ramps.

The F-15 Eagle has been deployed on a vast array of military operations including Iraq, Afghanistan, and Libya, to name but a few. During these operations, the F-15E has carried out deep strikes against high-value targets, combat air patrols, and provided close air support for coalition troops.

The McDonnell Douglas F-15 was to replace the McDonnell Douglas F-4 Phantom IIs and testing began in 1981 and ran until 1983, during which the F-15E logged more than 200 flights, demonstrated take-off weight of more than 70,000 pounds, and tested over sixteen different weapons-carrying configurations.

In 1984, the USAF chose the F-15 over the Panavia Tornado and General Dynamics F-16, believing that the F-15E had more potential to grow. As a result, the USAF purchased over 392 aircraft. The initial versions of the F-15s were the F-15 single-seater and the TF-15 twin-seater, and the first F-15A flight was made in 1972 with the first flight of the twin-seater F-15B flown in 1973.

The F-15C single-seater and the F-15D twin-seater models entered production in 1978 and conducted their first flights in that year. The Boeing F-15E strike variant was selected for production over General Dynamics' competing F-16XL in 1984.

In 2007, 178 USAF F-15Cs were upgraded with the AN/APG-63(V)3 Active Electronically Scanned Array radar, and a considerable number of F-15s are to be equipped with the Joint Helmet Mounted Cueing System (JHMCS). Lockheed Martin is working on an IRST system for the F-15C.

The F-15 has an all-metal semi-monocoque fuselage with a large cantilever shoulder-mounted wing. The empennage is metal and of composite construction, with twin aluminium/composite material honeycomb structure fins with boron-composite skin, resulting in an exceptionally thin tail plane and rudders with all-moving composite horizontal tail surfaces outboard of

F-15C 84-0001 from the 493rd Fighter Squadron, 'Grim Reapers', departs RAF Lakenheath with full reheat.

Above: F-15E 98-0131 from the 492nd Fighter Squadron, 'Mad Hatters', on the 'Last Chance' being disarmed by ground personnel before heading out on a night sortie.

Below: F-15E 00-3001 from the 494th Fighter Squadron, 'Panthers', on short finals with air brakes applied at RAF Lakenheath.

the fins. The F-15 has a spine-mounted air brake and retractable tricycle landing gear. It is powered by two Pratt & Whitney F100 Axial compressor turbofan engines, with afterburners mounted side-by-side in the fuselage, fed by intake ramps. The cockpit is mounted high in the forward fuselage with a one-piece windscreen and large canopy to increase visibility.

The F-15's manoeuvrability is derived from its low wing loading, giving a high thrust-to-weight ratio, enabling the aircraft to turn tightly without losing airspeed. The F-15 can climb to 30,000 feet in around sixty seconds. The thrust output of the dual engines is greater than the aircraft's weight, thus giving it the ability to accelerate in a vertical climb, and the F-15s design allows the weapons and flight controls to be safely and effectively operated by one person during air to air combat.

The 'A' and 'C' variants are single-seat and provide the main air superiority, while the 'B' and 'D' models are the two-seat variants used for training purposes; the 'E' model is also a two-seater, but with a weapon systems officer.

A multi-mission avionics system includes a HUD, advance radar with inertial guidance system (INS), flight instruments, ultra-high frequency (UHF) communications, tactical air navigation (TACAN), and instrument landing system (ILS) receivers. It also has an internally mounted, tactical electronic warfare system, identification friend or foe (IFF) system, electronic countermeasures suite, and a central digital computer.

The HUD provides essential flight information; this display, visible in any light condition, provides the pilot with the information necessary to track and destroy an enemy aircraft without having to look down at cockpit instruments.

The APG63 and APG70 Pulse Doppler radar system can look up at high-flying targets and look down, shooting at low-flying targets without being confused with ground clutter.

The APG radar can detect and track aircraft and high-speed targets at distances from close range to beyond visual range. The APG-63 has a basic range of 100 miles and feeds target information into the central computer for effective weapons delivery. For close-in dogfights, the radar automatically acquires enemy aircraft, and this information is projected on the HUD. The electronic warfare system provides both threat warning (radar warning receiver) and automated countermeasures against selected threats.

The F-15 is equipped with an automated weapon system, enabling the pilot to release weapons effectively, while being able safely use the HUD, avionics, and weapon controls located on the engine throttles or control stick.

The F-15 can also be equipped with a combination of weapons, which include AIM-7F/M Sparrow missiles or AIM-120 AMRAAM advanced medium-range air-to-air missiles on its lower fuselage corners, AIM-9L/M Sidewinder or AIM-120 AMRAAM missiles on two pylons under the wings, and an internal 20-mm M61 Vulcan Gatling gun in the right-wing root.

The McDonnell Douglas F-15E Strike Eagle has two seats and is a dual-role integrated fighter for all-weather, air-to-air, and deep interdiction missions. For low-altitude, high-speed penetration and precision attack on tactical targets at night or in adverse weather, the F-15E carries a high-resolution APG-70 radar and LANTIRN pods to provide thermography.

The largest operator of the F-15 is the USAF. The first Eagle (F-15B) was delivered in 1974 and the first for use by a combat-ready squadron in 1976. The first aircraft was shot down by an Israeli Air Force F-15 in 1979, during Israeli raids against Palestinian factions in Lebanon. It was reported that the F-15A shot down thirteen Syrian MIG-21 and two Syrian MIG-25.

The USAF began deploying F-15C, 'D', and 'E' variants to the Persian Gulf region in 1990 for Operations Desert Shield and Desert Storm during the Gulf War combat against Iraqi forces. The F-15 accounted for thirty-six of the thirty-nine air-to-air victories by the USAF, and Iraq confirmed the loss of twenty-three of its aircraft in air-to-air combat.

The F-15C and 'D' fighters were used in the air superiority role, while the F-15E Strike Eagles were used in air-to-ground attacks mainly at night, hunting modified Scud missile launchers and artillery sites using the LANTIRN system. The USAF reported that the F-15Cs had thirty-four confirmed kills of Iraqi aircraft during the 1991 Gulf War, most of them by missile fire, which included five MiG-29s, two MiG-25s, eight MiG-23s, two MiG-21s, two Su-25s, four Su-22s, one Su-7s, and six Mirage F1s, plus a few other aircraft.

Amazingly, air superiority was achieved within the first three days of the conflict, while many of the later kills were reportedly of Iraqi aircraft with pilots fleeing to Iran rather than engaging USAF aircraft. A F-15E achieved an aerial kill of an Iraqi Mi-8 helicopter with a laser-guided bomb, while two F-15Es were lost to ground fire and another was damaged on the ground by a Scud strike on Dhahran air base.

In 2009, the last F-15A from the Oregon Air National Guard was retired, which also included the F-15B variant in the US. The F-15C and 'D' variants will be retiring soon to make way for the developed F 22 Raptor, but due to production problems, the F-15 will continue longer than they expected.

As of 2013, regular USAF F-15C and F-15D fighters are based overseas with the Pacific Air Forces at Kadena Air Base in Japan and with the USAFE at RAF Lakenheath in the UK.

The UAF plans to keep over 175 F-15C/Ds along with over 220 F-15Es in service beyond 2025. The F-15E will remain in service for years to come because of the model's primary air-to-ground role and the lower number of hours on the F-15E airframes. As part of the USAF retirements in response to sequestration, unfortunately over 250 F-15C will retire to eventually be replaced by the F-35 Lightning II and the F-22.

F-15C 86-0178 from the 493rd Fighter Squadron parked inside the HAS at RAF Lakenheath.

HH-60G Pave Hawk

The Sikorsky MH-60G/HH-60G Pave Hawk is a twin turboshaft engine helicopter in service with the USAF. It is a derivative of the UH-60 Black Hawk and incorporates the USAF PAVE electronic systems program. The HH-60/MH-60 is a member of the Sikorsky S-70 family.

The MH-60G Pave Hawk's primary mission is insertion of and recovery of special operations personnel, while the HH-60G Pave Hawk's core mission is recovery of personnel under stressful conditions, including search and rescue. Both versions conduct day and night operations into hostile environments. Because of its versatility, the HH-60G may also perform peace-time operations such as civil search and rescue, emergency aeromedical evacuation (MEDEVAC), disaster relief, international aid, and counter-drug activities.

In 1981, the USAF chose the UH-60A Black Hawk to replace its HH-3E 'Jolly Green Giant' helicopters. After acquiring some UH-60s, the USAF began upgrading each with an air refuelling probe and additional fuel tanks in the cabin. The machine guns are 0.50-inch (12.7-mm) XM218s. These helicopters were referred to as 'Credible Hawks' and entered service in 1987.

Afterwards, the Credible Hawks and new UH-60As were upgraded and redesignated MH-60G Pave Hawk. These upgrades were to be completed in a two-step process. However, funding restrictions only allowed sixteen Credible Hawks to receive the second step equipment. These helicopters were allocated to special operations use. The remaining eighty-two Credible Hawks received the first step upgrade equipment and were used for combat search and rescue. In 1991, the search and rescue Pave Hawks were redesignated HH-60G.

The Pave Hawk is a highly modified version of the Sikorsky UH-60 Black Hawk, which features an upgraded communications and navigation suite that includes an integrated inertial navigation/global positioning/Doppler navigation system, satellite communications, and secure voice. The term 'PAVE' stands for Precision Avionics Vectoring Equipment.

All Pave Hawk HH-60Gs have an automatic flight control system, night vision goggles lighting, and forward-looking infrared system that greatly enhances night-time low-level operations. Additionally, some Pave Hawks have colour weather radar and an engine/rotor blade anti-ice system that gives the HH-60G an all-weather capability. Pave Hawk mission equipment includes a retractable in-flight refuelling probe, internal auxiliary fuel tanks, two crew-served (or pilot-controlled) 7.62-mm mini-guns or .50-calibre machine guns and an 8,000-lb (3,600-kg) capacity cargo hook. To improve air transportability and shipboard operations, all HH-60Gs have folding rotor blades.

Pave Hawk combat enhancements include a radar warning receiver, infrared jammer, and a flare/chaff countermeasure dispensing system. HH-60G rescue equipment includes a hoist capable of lifting a 600-lb (270-kg) load from a hover height of 200 feet and some personnel locating system. Several Pave Hawks are equipped with an over-the-horizon tactical data receiver that can receive near real-time mission update information.

During Operation Desert Storm, Pave Hawks provided combat search and rescue coverage for coalition air forces in western Iraq, Saudi Arabia, coastal Kuwait, and the Persian Gulf. They also provided emergency evacuation coverage for US Navy SEAL teams penetrating the Kuwaiti coast before the invasion.

During Operation Allied Force, the Pave Hawk provided continuous combat search and rescue coverage for NATO air forces, and successfully recovered two USAF pilots who were isolated behind enemy lines.

In March 2000, three Pave Hawks deployed to Hoedspruit Air Force Base in South Africa to support international flood relief operations in Mozambique. The HH-60Gs flew 240 missions in seventeen days and delivered more than 160 tons of humanitarian relief supplies.

Pave Hawks have regularly operated during Operation Iraqi Freedom and Operation New Dawn, and continue to be operated in Operation Enduring Freedom, supporting Army and Marine Corps' ground combat operations and standby search and rescue support for US and Coalition fixed-wing combat aircraft supporting those ground operations.

Unfortunately, one of the 56th RSQ Pave Hawks crashed in 2014 during a routine low-flying training exercise, when it encountered a massive bird strike. The crash killed the pilots and crew of the trailing helicopter, and the heartache was felt across the whole of the local community and on both US Bases.

The Pave Hawk assisted with the search and rescue of a 494th Squadron F-15/D jet based at RAF Lakenheath that crashed into a field near Spalding, Lincolnshire. Thankfully, the aircrew survived.

The 56th and 57th Rescue Squadrons began relocating from RAF Lakenheath to Aviano Air Base, Italy, in 2017 as part of a series of moves outlined in the European Infrastructure Consolidation review positioning them closer to potential trouble spots in Eastern Europe and Africa.

Above: Sikorsky HH-60G Pave Hawk 89-26206 parked on the 56RQS dispersal at RAF Lakenheath.

Next page

Above: Sikorsky HH-60G Pave Hawk 89-26205 taxies out of the 56RQS dispersal and to the runway for a training sortie at RAF Lakenheath.

Middle: Sikorsky HH-60G Pave Hawk 89-26205 hovers while awaiting air traffic control instructions at RAF Lossiemouth during Exercise Joint Warrior.

Below: Sikorsky HH-60G Pave Hawk 89-26212 taxies back to its assigned parking ramp at RAF Lossiemouth while on Exercise Joint Warrior.

LN
USAF
26205

6212

8
A Day in the Life of a 48th Fighter Wing Pilot

A day in the life of a USAF F-15 Strike Eagle fighter pilot typically starts in the early hours of the morning with a mug of strong coffee and some popcorn. This is then followed by a mass briefing that everyone must attend, and which only lasts around thirty minutes. The flight leader ensures everyone is aware of the special instructions and safety rules governing the forthcoming flight.

Once the mass brief is complete, pilots go straight into the flight briefing, which will last about two hours. During this, pilots receive vital flight information and procedures, which include pre-flight checks, ground ops, take off, departure, airspace checks, and post-flight flows.

The flight briefing ends, granting aircrew a chance to have a relaxing period before their mission. Thirty minutes pass, the aircrew are dressed in their flight gear, they all meet at the squadron operations desk known as 'top 3', which consist of the squadron commander, director of operations, and squadron supervisor.

The squadron operations supervisor gives the final briefing, which outlines any essential weather forecast for their mission and an up-to-date airfield status. The supervisor also tells the aircrew which jet they are each assigned to and which Permanent Aircraft Shelter (PAS) the jet is parked.

Normally the journey to the jet is a short trip taken by a crew bus (though some aircrew opt to walk depending on the weather and how far away the jet is parked). Once at the jet, the crew chief salutes and takes the helmet bag and publications kit containing maps and approach plates from the pilot. The crew chief follows the pilot as they go through their pre-flight inspection of the aircraft.

The pilot is now seated and strapped in and going through his visual and pre-flight checks while in communications with his crew chief, flight leader, and ground controller. With the canopy lowered, 'Tacho' and 'Hitman' flight is ready for engine start.

With all systems good, the pilot has been given clearance to taxi from air traffic control and the aircraft taxi out from their individual hangars and head towards the main runway.

While the aircraft taxi, the pilot will be running through essential pre-flight instrumental checks, plus talking to the aircrew to make sure everything is OK before stopping at the arming area (EOR).

Ground personnel give the jets a final visual inspection before taxi for take-off clearance is requested and given. 'Tacho' flight is a pair of F-15c, which is an upgrade for the pilots.

The aircraft will head to the north of the UK, and as for Hitman flight, again a pair of F-15c, it will be heading for the low-level corridor and Low Flying Area 7 (LFA7), also known as Mach Loop.

The Mach Loop, known to the locals as the Machynlleth Loop, is a selection of Valleys in west–central Wales. The region is used as a low-level training area for front-line fast jet and propeller aircraft. The series of valleys lies 8 miles east of Barmouth and is nestled between the towns of Dolgellau to the north and Machynlleth to the south.

The training area is within Low Flying Area (LFA) 7, which covers most of Wales, and aircraft that use the training area include RAF Tornados, Typhoons, and Hawks, and C130 Hercules as well as USAF F-15 Eagles, which are based at RAF Lakenheath.

Low-flying remains an essential skill for military aircrew. It allows them to undertake various roles and manoeuvres, flying as low as 250 feet from the nearest terrain, and aircraft can be seen flying through Wales in all kinds of weather scenarios. Low-level flying requires continuous practice and training as it is a highly demanding skill for all pilots. As we live in an uncertain world, it is important that our armed forces are fully trained so that if they are ever called upon to protect the nation or provide humanitarian aid, they will be ready.

Both sets of aircraft have been given taxi clearance for runway 24 at RAF Lakenheath with 'Tacho' flight in front followed closely by 'Hitman' flight. Prior to any aircraft taking off, the flight leader will visually or verbally confirm that each aircraft in their flight has been armed and that before take-off checks have been completed.

Air Traffic Control: 'Tacho flight, cleared to take off runway 24, winds 230 at 4 knots, after departure climb flight level 10 contact departure 250.300'.

'Tacho' flight starts their take-off roll, engaging after-burner. Within seconds, it rotates off the runway: '"Tacho" flight airborne passing two thousand, Deconfliction/Traffic Service'. 'Tacho' flight now heads north on its upgrade mission while 'Hitman' flight has already lifted off the runway and is climbing to 10,000 feet. '"Hitman" flight airborne passing two thousand, Deconfliction/Traffic Service', RAPCON controllers give further instructions for hitman flight to climb to medium level normally 19,000 feet before being handed over to Swanwick control. Swanwick control would guide 'Hitman' flight through narrow gaps in the civilian airways, one popular corridor known as the Lichfield Corridor. Once approaching the low-level area, 'Hitman' flight would request decent from air traffic. '"Hitman" flight of three requests decent to low level', Swanwick would respond: '"Hitman" flight of three taking your own terrain clearance, decent is approved'.

'Hitman' flies off to join the low-level area and descend to 250 feet, flying through the valleys of Wales. It passes through the Cad Pass and goes around the loop two times, giving the aviation photographer great photographic opportunities before the flight leader calls 'knock it off'. It then breaks out and rejoins controlled airspace by contacting Swanwick control for the return journey back to RAF Lakenheath.

Sortie durations vary depending on the mission, typically averaging one hour and thirty minutes from take-off to returning to base.

Once Swanwick have handed 'Hitman' flight back to controllers at RAF Lakenheath, aircraft will rejoin the pattern via Bravo and Charlie waypoints, which are north of the airfield, or Delta, which is south. Once pilots have established communication with controllers, they will be directed into the pattern for either pattern work or a full stop landing.

Above left: Aircrew apply all their flight gear ready for their flight.

Above right: Pilot's Helmet 'Torc' from the 493rd Fighter Squadron, RAF Lakenheath.

Aircrew gather at the mission's operation desk for their final briefing before heading out to the aircraft.

Capt. A. 'Kamikazze' Eshkenazi gathers his thoughts before taking his seat in the F-15C 86-0178 from the 493rd Fighter Squadron.

Above: Capt. M. 'Switcha' Rioux conducts his visual checks of the F-15C 86-0164 from the 493rd Fighter Squadron.

Below: Capt. Rioux completes his checks of the F-15C 86-0164 and begins to climb into the aircraft.

The crew chief conducts visual checks of the F-15C and communicates with the aircraft commander.

Above: Capt. Eshkenazi communicating with his crew chief before requesting engine start.

Below: Capt. Eshkenazi checking his instruments and the crew chief conducts visual checks.

Capt. Eshkenazi communicating with his ground crew while parked inside a PAS.

Above: Capt. Eshkenazi conducts final checks before he begins the engine start up process.

Below: Final checks complete, the ground crew remove covered to prepare engine start up.

Above left: Engine starts approved, the canopy is lowered, and the crew chief looks on.

Above right: Capt. Eshkenazi conducts his engine start up procedure, making sure all system are good to go.

Above left: Capt. Eshkenazi releases the brakes and applies a little power and the F-15C begins to taxi out of the HAS.

Above right: With Air Traffic Control giving clearance to taxi, the aircraft commander begins to taxi to the end of the runway hold point (EOR) or 'Last Chance'.

Above left: F-15E Strike Eagle 98-0132 from the 494th Fighter Squadron taxies towards the EOR.

Above right: F-15E Strike Eagle 01-2003 from the 494th Fighter Squadron taxies towards the EOR during an evening sortie.

F-15E Parked at the end of the runway holding point, known as EOR. Ground crew conduct final checks of the aircraft before they can launch.

Ground crew conduct visual checks of the F-15E before it is given clearance for departure.

F-15C 01-0166 at the EOR while ground crew conduct visual checks on the aircraft.

Above: Crews await Air Traffic Control clearance while waiting at the EOR at RAF Lakenheath.

Below: Three F-15E Strike Eagles from the 494th Fighter Squadron waiting further instructions from Air Traffic Control while at the EOR.

Above left: Ground crew disarm F-15C 86-0169 from the 493rd Fighter Squadron at the EOR.

Above right: A pair of 493rd Fighter Squadron F-15Cs hold at the EOR holding point waiting for further instructions during a night sortie at RAF Lakenheath.

Above left: Capt. Rioux in F-15C 86-0164 is given taxi clearance by Air Traffic Control, with his five colleagues following behind.

Above right: Capt. taxies to RAF Lakenheath active runway 24.

Below left: Capt. Eshkanazi holds at the EOR, waiting for Air Traffic Control to give further instructions.

Below right: The aircraft commander communicates with Air Traffic Control while holding at the EOR at RAF Lakenheath.

Three F-15C from the 493rd Fighter Squadron holding at the EOR while ground crew conduct final checks.

Above: Capt. Eshkanazi going through final checks in F-15C 86-0178 at the EOR.

Below: Capt. Rioux in F-15C 86-0164, waiting on further instructions from Air Traffic Control while parked at the EOR.

Capt. Eshkanazi taxies to the active runway at RAF Lakenheath in F-15C 86-0178.

Above: F-15C 86-0178, call sign 'Tacho' flight with full reheat takes off from RAF Lakenheath active runway 24 on an upgrade mission.

Below: F-15C 86-0164 from the 493rd Fighter Squadron, 'Grim Reapers', takes off from runway 24 at RAF Lakenheath.

Above: Capt. Eshkanazi in F-15C 86-0178 with full reheat departs RAF Lakenheath on an upgrade flight with call sign 'Tacho'.

Right: F-15C 84-0044, call sign 'Hitman', is holding on the northern side of the runway awaiting, its slot to depart RAF Lakenheath.

Below: F-15E Strike Eagle 91-0302 from the 492nd Fighter Squadron departs with full reheat on another training mission.

Above left: F-15C 86-0171, call sign 'Tacho' flight, departs with full reheat from RAF Lakenheath on an upgrade flight with the control tower in the background.

Above right: F-15C 86-0159, call sign 'Hitman', engaged full reheat and is at rotation speed and lifts off out of RAF Lakenheath.

Above left: F-15C 86-0159, call sign 'Hitman', with full reheat takes off out of RAF Lakenheath to join up with the rest of 'Hitman' flight.

Above right: F-15C 84-0027 from the 493rd Fighter Squadron, 'Grim Reapers', flying low level through the Welsh Mountains—this route is known as the 'Mach Loop'.

Below left: F-15C 84-0010 from the 493rd Fighter Squadron turns left at low level altitude towards Cad East in the Welsh mountain ranges.

Below right: F-15C 84-0010 low level while flying through 'Mach Loop' passing the Bwlch mountain range.

F-15C 86-0164 from the 493rd Fighter Squadron banks hard left from the exit to show the topside profile of the F-15C.

Above: F-15C 84-0027 making his final low-level pass through the Welsh mountains before heading back to RAF Lakenheath.

Below: A F-15E flying low level through the mountains in Wales, LFA7, better known as the 'Mach Loop'.

F-15E 91-0324 from the 494th Fighter Squadron flying low level through the Bwlch, known as 'Mach Loop' or LFA7.

Above: 'Harley' flight begin their low-level transit through LFA7—better known as the 'Mach Loop'.

Below: F-15E 00-3001 from the 494th Fighter Squadron, 'Panthers', extremely fast and low through LFA7.

'Harley' flight pull some wing vapour as they begin their low-level transit through the 'Mach Loop'.

Above: F-15E 97-0220 from the 492nd Fighter Squadron flies over the top of LFA7.

Below: F-15E 91-0314 from the 494th Fighter Squadron knocks off the low-level training and heads back to RAF Lakenheath.

Once back on the ground, the pilot stops at the rearming area, known as the end of runway (EOR), before heading back to the parking areas and going through maintenance debrief with the crew chief.

Aircrew then return to the squadron building where the aircrew pack away their flight gear and tell the squadron operation supervisor the maintenance status of the aircraft. Then they will find their flight leader to establish a debrief time.

Once the mission is completed, it can take around four hours to complete a debrief as every part of the mission is recorded; during this, the aircrew are able to look at each display and listen to the radio calls before heading into their mass briefing. This consists of assessing each simulated missile shot to ensure it is a valid shot against a target that has been correctly identified as 'hostile', as they relate to the overall objectives of the mission.

Besides flying, which has virtually filled the day already, aircrew also have a 'real job'. There are a multitude of jobs in the squadron—standardisation/evaluation, training, weapons, life support, and safety are the most common.

Virtually every pilot has a similar workload, and the workload generally increases with rank and responsibility. That is why time management and teamwork are vital within the squadron. The scheduling and training departments normally work two weeks in advance. They have three officers working to help allocate aircrew that are due normal and upgrade training, and who make sure the pilots are scheduled for the correct training syllabus.

This would include training for a new pilot to the squadron, making sure they are familiar with UK procedures and squadron procedures and to establish squadron pilots going for command upgrades. Normally, a pilot will fly one event a day as they will have additional duties to carry out once they return to squadron, which traditionally will take another two to three hours to complete. The 493rd Squadron's aircraft activity normally depends on the training needs of the pilots, but the traffic flow normally sees ten jets fly in morning and eight in the afternoon sorties.

Every pilot must also spend time in the flight simulator, and at RAF Lakenheath, they have four 360-degree simulators that can all be connected and linked up to other simulators in any US airbase back in the States. This helps pilots coordinate for future mission and ongoing training programs. A pilot must fly eight to nine times a month, plus two to three simulator missions, to stay fully mission combat ready.

Training to become an USAF F-15 pilot starts with the training academy for a year and half, achieving all the relevant grades before being selected for the F-15s. Full training begins on the Northrop T-38 Talon for three months (around seventy hours) before moving onto the F-15s, where more training is given. Once a pilot has graduated, they will be assigned to a squadron six to eight months later.

The pilot call sign is either pre-allocated by air traffic control or pre-assigned by the squadron on a selection of call signs that the squadron can use. As a tradition, the squadron have a call sign auction, where pilots buy their call signs—the best go for more money. The best call signs include 'Death', 'Bullet', and 'Deadly'. Once the pilot has paid for a certain call sign, it is theirs for the time they are in the squadron. The aircrew who owes money to the squadron would get a funny call sign of 'Licker'. Some call signs must be used during operational flights, which include the two-ship upgrade 'Tacho', squadron commander 'Reaper', director of operations 'Grim', and weapons officer 'Rambo'.

Capt. Eshkanazi in F-15C 86-0178 breaks left for the downwind leg and full stop landing runway 24 at RAF Lakenheath.

Above: F-15E 91-0344 from the 494th Fighter Squadron breaks to the left for a full stop landing at RAF Lakenheath.

Below: F-15C 84-0027 *MIG Killer* from the 493rd Fighter Squadron landing at RAF Lakenheath.

Above left: F-15E from the 492nd on finals for runway 24 at RAF Lakenheath.

Above right: Arriving back to RAF Lakenheath after another successful training sortie.

Above left: F-15E over the threshold of runway 24 at RAF Lakenheath.

Above right: F-15C 86-0175 from the 493rd parks outside the PAS while ground personnel check the aircraft.

Below left: The aircraft commander communicates with ground crew confirming the aircraft has been disarmed.

Below right: The crew chief disarming the F-15C after its training sortie.

Above left: The aircraft commander looks on while awaiting Air Traffic Control clearance to taxi back to his HAS.

Above right: F-15C 86-0159 from the 493rd Fighter Squadron taxies back to its PAS after a training mission.

Above left: F-15C 86-0172 from the 493rd Fighter Squadron taxies to its PAS after a mission.

Above right: The aircraft commander taxies his aircraft to his HAS.

Below left: The crew chiefs give the signal for Capt. 'Steal' Davis to apply the aircraft breaks while ground crew set the aircraft chocks.

Below right: The aircraft commander begins the engine shut down procedure.

Capt. Davis disembarks his aircraft after a training sortie.

The squadron schedule board, with all the pilots' relevant training and upgrade needs.

On non-flying days, all squadron pilots undertake other duties, which would include going to the gym for physical fitness training as aircrew must be in optimal health to handle the physical demands of modern aerial warfare. Aircrew as part of the squadron will have other demanding responsibilities working with squadron commanders to develop plans and implement policies; this could include new training or operational procedures. Other duties would include squadron scheduling, maintaining flight records, squadron life support, squadron training, and aircrew training devices. These varied functions monitor the status of aircrew training, document aircrew flight status, and provide training and equipment to aircrew.

The squadron structure normally is squadron commander, director of operations commander, squadron supervisor, flight commander, and WSO scheduling, training, and flight time. Flight commander responsibility is to support flying operations, coordinate aircraft issues, airspace coordination, mission briefing, and daily decisions due to weather, pilots, etc.

The aircrew standardisation/evaluation program provides commanders with a tool to validate individual aircrew along with their capabilities and qualifications. The program would also look at mission readiness and the effectiveness of unit flying. It would help plan the forthcoming training syllabus and scheduling, which aircrew on non-flying days would implement.

The standardisation flight regulates how the 48th conduct missions through published standard. Top gun competitions, weapons, and tactics ensure that the 48th FW is always achieving its mission goal. One of the most important programs the flight oversees is strike aircrew certification. The aircrew are grilled and constantly scrutinised to ensure they are mission ready all the time.

The standardisation and evaluation program publish the flying standards that all aircrew follow, and it ensures all flyers in the wing meet the demanding criteria. A formal check flight will take place every eighteen months and it is vital that quality assurance is also checked for all the maintainers in the wing. If an aircraft has undergone a major maintenance overhaul, the 48th Standardisation Flight will manage the functional check flight program in the wing that scrutinises the work performed on an aircraft that has received repair and maintenance. Weapons standardisation ensures that all weapons loaders in the wing meet a demanding level of performance. The advanced programs office is responsible for future hardware and software upgrades to all the wings F-15s.

The day of a pilot, whether flying on non-flying, is a very demanding role, but the rewards are outstanding. A normal day for a pilot is twelve hours on shift, but crew rest is also twelve hours. A normal assignment for a pilot on a squadron if he is single will be two years, but this can be extended to three years. If the pilot is married and the family is living with the pilot, it will be three years before they will be assigned to another wing.

Above left: F-15C 84-0015 parked inside a HAs awaiting its aircraft commander.

Above right: The 'Grim Reaper' F-15C is peacefully waiting its aircraft commander

Above: As the evening ends, the PAS lights shine on F-15D 84-0046.

Below: Capt. Eshkanai carries out his over duties back at the squadron building, while ground crew prep the aircraft for its next sortie.

9
Air Traffic Controlling at RAF Lakenheath and RAF Mildenhall

Air Traffic Control is a sophisticated profession and it takes a lot of demanding work and training over many years to become a competent controller. Within the USAF, the diverse aircraft movements from both RAF Lakenheath and RAF Mildenhall require controllers to be extremely professional while controlling many aircraft at any given time.

Air Traffic Controllers specialise in ensuring the safety and efficiency of air traffic, which involves establishing a sequence for the pilots to arrive and depart in a safe and orderly flow from both bases. All controllers, while based in the UK and Europe, must obey the Federal Aviation Authority (FAA) and the International Civil Aviation Organization (ICAO) plus Air Force Instructions (AFI) documentation.

The controllers act as the communication hub to the airfield, with communications involving everyone on the base with lines direct to the fire department, medical facilities, and base ops. Due to the very close vicinity of RAF Lakenheath and RAF Mildenhall, both sets of controllers will be in very close communication with each other. RAF Lakenheath has the main control with Radar Approach Control (RAPCON).

The personnel on both USAF airbases are all part of a complex network. Each has their role and duty towards the safety, upkeep, and flow of aircraft to and from both bases. It is a continual work flow. Maintaining aircraft, airbase duties, and air traffic safety are all paramount. Whether it is a live active operation or during peacetime duties, they are always at the ready.

Airfield operation is the heart of both bases. They are responsible for all aircraft entering the sensitive airspace that is classified as a Military Air Traffic Zone.

RAPCON (Radar Approach Control) are the eyes of the USAF and they control all aircraft movement on and around RAF Lakenheath. They also provide a radar support service to many civilian aircraft that are flying within the East Anglian area. RAPCON's job is to make sure all aircraft, both military and civilian, are deconflicted by coordinating their movements away from one another, helping them fly safely in the local area.

Centrally based is the control tower, and the personnel there monitor and control all aircraft movement on the base and all airborne aircraft that are considered within their visual traffic pattern. Other key staff members include the base operations section, who are key to both Lakenheath and Mildenhall. They have several roles to cover that include daily inspections and maintain the entire airfield infrastructure. This ensures they have continuous safe operational flying conditions always.

RAPCON are the eyes at RAF Lakenheath and RAF Mildenhall.

During normal training operations, controllers can control more than 70,000 take-offs, departures, and recoveries per year. With staffing as low as 62 per cent at times, airfield operations' can-do attitude shows a good example of the 48th Operation Support.

Both airfields have separate procedures to follow; this is due to the different aircraft types flown from the bases. RAF Mildenhall predominately operates with larger and slower aircraft, with the exception of the CV-22 Osprey.

RAF Mildenhall is operated twenty-four hours a day, seven days a week, but due to the built-up area of Mildenhall, the base has a Quiet Hour Operation from 11 p.m. until 7 a.m. local, but a waiver can be granted from commanders. Standard exceptions to Quiet Hour Operations are directed missions and aero-medical flights or transient arrivals and departures.

Aircraft either flying patterns or departing RAF Mildenhall will be advised by controllers to avoid flying over Mildenhall town, Barton Mills, Worlington, Tuddenham, Red Lodge, Beck Row, Isleham, Newmarket, West Row, Ely, and other towns and villages in the local area below 2,000 feet unless safety of flight is a factor.

Aircraft departing runway 29 or making approaches to runway 11 will be advised by controllers to avoid flying over Ely and Ely Cathedral below 2,000 feet unless safety of flight is a factor. Due to the close vicinity of houses and noise abatement procedures, aircraft are not authorised to perform unrestricted climbs at RAF Mildenhall.

The primary runway to be used at RAF Mildenhall is 29 when the wind speed is less than 5 knots, but the Tower Watch Supervisor (WS) will select the runway to be used. RAF Mildenhall Tower must coordinate runway changes with RAF Lakenheath before changing runways. RAF Mildenhall must also notify Airfield Management Operations (AM Ops), Base Weather, the fire department (FD), RAF Lakenheath RAPCON, and RAF Lakenheath Tower when the active runway direction changes. AM Ops shall notify Command Post (CP) and Bird Control of any active runway change. CP shall also notify the 100th Maintenance Operations Squadron, Maintenance Operations Centre, who then notifies Transient Alert of active runway change.

The standard pattern when using runway 29 is a left-hand pattern, and when runway 11 in use a right-hand pattern will be used. Air Traffic Control may direct different traffic patterns due to other traffic or airspace restrictions.

With normal pattern altitudes set at 3,000 feet mean sea level (MSL) and overhead pattern at 1,800 feet MSL, this must be coordinated by RAF Lakenheath RAPCON. The overhead pattern will be flown at 1,800 feet MSL until turning base leg. Aircraft wishing to descend below 1,800 feet MSL following the break turn must request and receive approval from the Tower.

RAF Mildenhall must coordinate through RAF Lakenheath RAPCON for aircraft arriving and departing RAF Mildenhall. RAF Lakenheath (RAPCON) will coordinate with London Military control and Swanwick Control.

RAF Mildenhall with RAPCON approval may authorise the use of specific departure and arrival routes into RAF Mildenhall, and Mildenhall Tower will transfer departing aircraft to RAPCON prior to exiting Mildenhall's combined military air traffic zone (CMATZ).

There are four VFR main departure and arrival routes for RAF Mildenhall, each named for the direction the aircraft will depart or arrive at RAF Mildenhall: northeast, northwest, southeast, and southwest. The northeast and northwest departure and arrival routes will not normally be approved during daytime hours or any night that the RAF Lakenheath traffic pattern has numerous departures or recoveries.

RAPCON will coordinate arriving aircraft to RAF Mildenhall tower prior to reaching 10 flying miles from the airbase. Arriving pilots would have to contact RAPCON 5 miles prior to any of the four entry points (CHETT, BARN, RISBY, and FEN), and would have been advised by RAPCON of their arrival intentions. If approved, the arriving aircraft would then fly through the appropriate entry point and proceed with recovery to RAF Mildenhall.

Normal phraseology for Air Traffic Controllers to use at RAF Mildenhall for departing aircraft would be as follows:

> Call sign: Depart Mildenhall via Radar Vectors, On Departure, Fly runway heading. Climb to Flight Level 100; expect further clearance once airborne, Departure Frequency 250.3, and Advice Ten minutes prior to departure for coordination of squawk.

All aircraft that wish to depart RAF Mildenhall must file a flight plan, which helps reduce aircraft departure as European air space is very congested, especially during the summer and school holiday periods. Central Flow Management Unit (CFMU) in Brussels, Belgium, frequently issues a Calculated Take-Off Time (CTOT) to an aircraft departing on a General Air Traffic flight plans. Normally, the calculated take-off time determines the time when the aircraft must take-off. Once a CTOT is issued, the aircraft must take off no earlier than five minutes prior, and no later than ten minutes after the issued CTOT time. AM Ops will notify Tower and RAPCON of all CTOTs that are issued. AM Ops relay the information to RAF Mildenhall tower and RAPCON at RAF Lakenheath on all proposed departure and inbound aircraft.

RAF Lakenheath Tower Air Traffic Controllers predominately operate and support the 48th FW's three combat ready F-15 fighter squadrons. The tower is also the main control hub with coordination with RAF Mildenhall, this is because RAF Lakenheath operates RAPCON. The role of RAPCON is a radar flight that provides information and advisories to arriving, departing, and *en route* aircraft. It also monitors flight progress of aircraft and receives, posts, and relays progress reports.

The radar position's priority is to ensure aircraft separation is kept under ICAO requirements plus initiate control instructions that includes monitoring and operating radio communications. This position also assists coordinates with all other radar departments. A radar assistant is there to aid the radar controller by accepting handoffs as directed by the radar positions. This helps continue a smooth operation of the sector and ensures the radar position is aware of any situations arising.

The feeder controller provides for the radar position sequencing and separation of aircraft vectored to intercept the final approach courses for handoff to the final controller. The final controller issues instructions to pilots based on the final approach course and glide path to the airfield.

RAF Lakenheath's normal operating hours are from 12 a.m. local time on Monday through to the end of local flying on Fridays; local flying is defined as flying in the UK, the North Sea, and the Dutch coast.

Aircrew must call the weather station to receive a finalised flight brief before their departure. Tower Watch Supervisor will notify 48th Operation Support Squadron (OSS) of any runway changes and the (OSS) will coordinate with the tower and RAPCON watch supervisor before a runway change is confirmed.

All flights will notify ground control when they are ten minutes prior to take-off. At this time, the flight commander will notify ground control if a formation take-off is planned and the flight leaders will remain on ground frequency until receiving their squawk. All flight members will acknowledge upon receipt. Before take-off, flight leaders will contact Squadron Operations for final mission updates prior to changing the flight to discrete frequency.

Once aircraft have been given taxi clearance, they will taxi using a clockwise direction to the end of runway (EOR), unless the ground controller states otherwise—the maximum taxi speed is 25 knots.

Once reaching the end of the runway, aircraft will take the first available arming spot flowing from the taxiway and exit the arming area by turning towards the runway. Aircraft loaded with live weapons will be armed and de-armed in the south arming area; they can also use an additional area to the north of the EOR at runway 24 when required, but when aircraft are using runway 06, there are currently no approved arm/de-arm EOR locations.

Once the aircraft have been armed and the ground chief technicians have completed their checks, the pilots will call air traffic control for take-off clearance. Once this has been approved, aircraft will begin their taxi run to the active runway. The minimum weather for take-off for the F-15s at RAF Lakenheath is 300-foot ceiling cloud base.

Prior to any aircraft taking off, the flight leader will visually or verbally confirm that each aircraft in their flight has been armed and that before take-off checks have been completed. No formation or maximum performance take-offs are permitted at RAF Lakenheath unless prior approvals granted by commanders. After take-off, aircraft will notify Departure Control with 'call sign, airborne passing two thousand, De-confliction/Traffic Service'.

Supersonic flights are not permitted over the UK unless authorised by commanders; however, supersonic flights are allowed when aircraft are heading directly out to sea and they may accelerate to supersonic speed when at least 10 NM out to sea and as low as 2,000 feet.

On returning to RAF Lakenheath, aircraft will rejoin the pattern via Bravo and Charlie, which are north of the airfield, or Delta, which is south. Once pilots have established communication with controllers, they will be directed into the pattern for either pattern work or a full stop landing.

Once the aircraft has taxied off the active runway and has notified the base of their location and intentions, aircraft must take the first available disarm location. Once disarm has been completed, aircraft can taxi back to the permanent aircraft shelter (PAS).

10
The 501st Combat Support Wing: RAF Alconbury

The 501st Combat Support Wing (501 CSW) represents the USAFE. Based in the UK, the unit operates from RAF Alconbury, a support base for RAF Fairford. Today, the 501st CSW ensures UK-based air base groups have the resources to perform well and are fully trained, supported, and equipped. This is to help them handle the highest command standards required. The unit will in turn be able to provide mission support, allowing US and NATO fighters to conduct a full scope of flying operations. The group covers military deployments or munitions movements, including any global command situations.

The 501st CSW currently supervises and supports four UK air base groups. It operates over seven units located in the UK and Norway: 421st Air Base Squadron at RAF Menwith Hill; 422nd Air Base Group at RAF Croughton, RAF Fairford, and RAF Welford; and 423rd Air Base Group at RAF Alconbury, RAF Molesworth, and Stavanger Air Base, Norway.

The wing has over 2,500 personnel. These are from both US military and civilian backgrounds, including over 180 MOD police personnel.

The 501st Bombardment Group (BG) was established in 1944 as a Very Heavy Bombardment Group flying B-29 Superfortress. In the summer that year, the group was sent to training station at Harvard Army Airfield, Nebraska. Due to a shortfall of B-29s, the group was re-equipped with former Bomber Command B-17 Flying Fortresses, which were previously used for training heavy bomber replacement personnel.

In 1945, the 501st BG was deployed to Pacific Theatre of Operations, assigned to the XXI Bomber Command, 315th Bombardment Wing, in the Northern Mariana Islands, Northwest Field, Guam. Due to a lack of facilities, on arrival at Guam, the group's personnel began to construct huts, and by early June 1945, most of the group's personnel could move into metal huts. After initial crew training, the group finally flew their first combat mission in late June 1945. Later, the missions were based on strategic attacks on enemy petroleum industries. These were very effective and hindered the enemy's ability to attack US forces. These attacks were so successful that the group received the Distinguished Unit Citation award for attacks on the Maruzen oil refinery at Shimotsu, the Utsubo oil refinery at Yokkaichi, and the petroleum centre at Kawasaki.

The 501st BG crews flew fifteen combat missions. The majority were dangerous operations that were conducted under the cover of night. The strategy was to fly its aircraft in a staggered formation at various altitudes *en route* to the target. The results were devastating, and this

type of attack was proved to be extremely successful. Combat operations ended because of the Japanese Capitulation in August 1945.

After the Victory over Japan Day, the 501st's mission was to drop supplies to Allied prisoners and to fly over Japan to evaluate bombardment damage. By Christmas 1945, the group was split, with a few of the B-29s flown back to the States for mothballing while the rest of the B-29s were left in service. Decisions were made, and the US started to consolidate various groups of the 315th BW in February 1946, with the 16th, 331st, and 502nd Groups being merged into the 501st. The other groups were deactivated in April in the same year. The 501st Bombardment Group was deactivated on 10 June 1946.

In August 1956, the 701st Tactical Missile Wing was established and finally activated on 15 September 1956. This unit was stationed at Hahn Air Base, West Germany; this was the first tactical missile wing in the USAF when activated on 18 June 1958, and it was then replaced by the 38th Missile Wing.

In 1975, the Soviet Union's development of the RSD-10 Pioneer missile gave major concerns for the NATO alliance. It was concluded that the striking power of the new missile if used would alter the security of Western Europe. The year 1980 saw the announcement by the US that it would deploy the General Dynamics BGM-109G Ground Launched Cruise Missile (GLCM) to Europe to counter this potential threat.

The wing was redesignated as the 501st Tactical Missile Wing (TMW) in January 1982; it was activated in July 1982 at RAF Greenham Common, Britain, to operate GLCMs. The first squadron of the 501st Tactical Missile Wing received its weapons in November 1983.

In 1986, during the months of August and September, a summit was held between the US and Russia. Some key points to be agreed were the removal of INF systems from Europe and equal global limits of 100 INF missile warheads, which led to the removal of all nuclear missiles from the base. This resulted in the 501st TMW being deactivated on 31 May 1991 as part of the withdrawal process.

The unit was redesignated the 501st Combat Support Wing (CSW) on 22 March 2005 and activated on 12 May 2005 at RAF Mildenhall, England. The 501st CSW's current mission is to manage and support various USAF units, installations, and activities in the UK, while not directly supporting operations at RAF Mildenhall or RAF Lakenheath. The 501st CSW was then relocated to RAF Alconbury and are still present there.

The 501st CSW provides and supports missions at the normally non-active RAF Fairford. Traditionally, the 501st CSW helps and supports the Royal International Air Tattoo, but recently the 501st has been active at RAF Fairford. The unit was part of the NATO-led exercises BALTOPS and Ample Strike.

The 421st Air Base Squadron

The 421st Air Base Squadron provides a variety of base and mission support, supplying communications and intelligence support services to the UK and the US National Security Agencies. The airbase contains an extensive satellite ground station and is a communications intercept and missile warning site that has been described as the largest electronic monitoring station in the world. With a community of more than 2,000 military, civilian, contractors, and dependents at RAF Menwith Hill, it ensures outstanding facilities and emergency, safety, legal, postal, and force support services to ensure superior quality of life and to enable and enhance mission success.

The 422nd Air Base Group

The 422nd Air Base Group is located at RAF Croughton, Northamptonshire. The 422nd provides world-class combat support to all installations, services, and force protection, which enables global correspondence and strike operations at RAF Croughton, RAF Fairford, and RAF Welford.

The 420th Air Base Squadron

The 420th Air Base Squadron provides exceptional mission support, acquiring, projecting, and sustaining deployed forces. This enables US and NATO personnel to co-ordinate a full variety of flying operating from RAF Fairford.

The 420th Munitions Squadron

The 420th Munitions Squadron provides immediate combat reserve materiel and performs routine munitions stockpile maintenance checks. These checks are key in supporting the USAFE's only bomber forward operating location at RAF Fairford.

The 422nd Air Base Squadron

The 422nd Air Base Squadron provides outstanding combat support and quality of life for over 5,000 US and UK personnel, delivering premier logistic, personnel, and service support. The 422nd Civil Engineer Squadron provides, manages, and preserves a sustainable installation through engineering programs and emergency response services.

The 422nd Communications Squadron

The 422nd Communications Squadron helps to establish, maintain, and employ cyberspace to enable the NCA, CCDR, and UK mission partners to achieve US national goals and objectives.

The 422nd Medical Squadron

The 422nd Medical Squadron helps provide innovative healthcare and ensures medical readiness for airmen, families, and mission partners. The unit also helps to enable the success of communications and global strike operations.

The 422nd Security Forces Squadron

The 422nd Security Forces Squadron defends all personnel and resources of the 422nd Air Base Group. This enables them to provide intelligence, communications, and global strike operations in support of US, NATO, and mission partner objectives worldwide.

The 423rd Air Base Group

The 423rd Air Base Group, based at RAF Alconbury, is composed of six squadrons that support over 6,300 personnel in the UK and over 1,200 in Norway.

The 423rd Civil Engineer Squadron

The 423rd Civil Engineer Squadron's energetic personnel consists of military, MOD, and Department of Defense civil servants. The 423rd Civil Engineer Squadron (CE) provides support for 501st CSW's missions at RAF Alconbury and RAF Molesworth. The 423rd CE oversees over 700 facilities and 950 acres of land. The unit also conducts facility construction, operations, and maintenance. In addition, it also provides environmental, housing, and emergency services for 6,500 personnel. The 423rd Civil Engineer Squadron supports eleven international units.

The 423rd Communications Squadron

The 423rd Communications Squadron provides leading communications support, securing a global connectivity for the US and its mission partners.

The 423rd Force Support Squadron

The 423rd Force Support Squadron provides support readiness and regeneration of airmen by providing exceptional family care, personnel support, and fitness. They also have quality of life programs that enhance morale for the RAF Alconbury and RAF Molesworth community.

The 423rd Medical Squadron

The 423rd Medical Squadron (MDS) is based at RAF Alconbury. The 423rd MDS clinic provides routine medical and dental care to all active duty and dependent personnel from the 501st CSW.

The 423rd Security Forces Squadron

The 423rd Security Forces Squadron safeguards personnel and dependants at both RAF Alconbury and RAF Molesworth. The 423rd Security Forces Squadron protects up to 6,000 military and civilian personnel who can gather intelligence for three installations. The unit also assists the 426th Air Base Squadron at Stavanger in Norway and can deploy worldwide to support national security.

The 426th Air Base Squadron

The 426th Air Base Squadron provides support with unique practical formats that delivers to over 250 US military personnel along with their dependents. Assigned to NATO, the 426th Air Base Squadron is also responsible for ensuring the movement and customs of all department of defence cargo supporting over $900 million war reserve materiel programs at thirteen sites in Norway.

11
Summary

The US presence will continue for the foreseeable future in the UK, even though the structure of the active squadrons, units, and air bases may change. The future of RAF Mildenhall has been extended now until 2020, and there have been rumours that RAF Fairford may be taking more of an active role, which would see bomber and transport deployments becoming a regular occurrence.

The continued active aggression from Russia and the rise of the Islamic State has thrown the US government's budgets into disarray. With a new president and government, the military budget is being re-evaluated and there are hopes that there will be more money spent on military assets and personnel.

The special relationship is an unofficial term for the political, diplomatic, cultural, economic, military, and historical relations between the UK and the US. It was used in a 1946 speech by Winston Churchill and continues at present with the new administration (President Trump) and Prime Minister Theresa May.

President John F. Kennedy's speech, bracing the nation for the Cuban Missile Crisis, stated on 22 October 1962:

> The path we have chosen for the present is full of hazards, as all paths are; but it is the one most consistent with our character and courage as a nation and our commitments around the world. The cost of freedom is always high, but Americans have always paid it. And one path we shall never choose, and that is the path of surrender or submission. Our goal is not the victory of might, but the vindication of right; not peace at the expense of freedom, but both peace and freedom, here in this hemisphere, and, we hope, around the world. God willing, that goal will be achieved.

Let us not forget these airmen, soldiers, and sailors who have paid the ultimate sacrifice in the line of duty. They all will be cared for throughout their final journey home. Remember your heritage and heroes, your family members, your loved ones, your neighbours, and your friends who have given the ultimate sacrifice or lost a loved one in defence of liberty's blessings.

The US cemetery in Cambridge, UK, contains the remains of 3,812 military dead.

The flagpole platform near the main entrance. The great mall with its reflecting pools stretches eastward.

Bibliography

en.wikipedia.org/wiki/MDP_Wethersfield
en.wikipedia.org/wiki/RAF_Alconbury
en.wikipedia.org/wiki/RAF_Barford_St_John
en.wikipedia.org/wiki/RAF_Bentwaters
en.wikipedia.org/wiki/RAF_Blenheim_Crescent
en.wikipedia.org/wiki/RAF_Burtonwood
en.wikipedia.org/wiki/RAF_Chelveston
en.wikipedia.org/wiki/RAF_Chicksands
en.wikipedia.org/wiki/RAF_Croughton
en.wikipedia.org/wiki/RAF_Fairford
en.wikipedia.org/wiki/RAF_Feltwell
en.wikipedia.org/wiki/RAF_Greenham_Common
en.wikipedia.org/wiki/RAF_Lakenheath
en.wikipedia.org/wiki/RAF_Little_Rissington
en.wikipedia.org/wiki/RAF_Manston
en.wikipedia.org/wiki/RAF_Menwith_Hill
en.wikipedia.org/wiki/RAF_Mildenhall
en.wikipedia.org/wiki/RAF_Molesworth
en.wikipedia.org/wiki/RAF_Sculthorpe
en.wikipedia.org/wiki/RAF_Shepherds_Grove
en.wikipedia.org/wiki/RAF_Upper_Heyford
en.wikipedia.org/wiki/RAF_Upwood
en.wikipedia.org/wiki/RAF_Welford
en.wikipedia.org/wiki/United_States_Air_Force_in_the_United_Kingdom
en.wikipedia.org/wiki/Woodbridge_Airfield
www.352sow.af.mil/
www.501csw.usafe.af.mil/
www.lakenheath.af.mil/
www.mildenhall.af.mil/